T0180194

Graphene

Tianrong Zhang

Graphene

From Theory to Applications

Guangxi Science & Technology
Publishing House

Springer

Tianrong Zhang
Chicago, IL, USA

ISBN 978-981-16-4591-4   ISBN 978-981-16-4589-1   (eBook)
https://doi.org/10.1007/978-981-16-4589-1

Jointly published with Guangxi Science & Technology Publishing House
The print edition is not for sale in China (Mainland). Customers from China (Mainland) please order the print book from: Guangxi Science & Technology Publishing House.
ISBN of the Co-Publisher's edition: 978-755-51-1227-3

Translation from the Chinese language edition: *Random Walk and Gecko Tape: Graphene Coming Out of the Ivory Tower* by Tianrong Zhang, © Guangxi Science & Technology Publishing House 2020. Published by Guangxi Science & Technology Publishing House. All Rights Reserved.

This Springer imprint is published by the registered company Springer Nature Singapore Pte Ltd.
The registered company address is: 152 Beach Road, #21-01/04 Gateway East, Singapore 189721, Singapore

Preface

Throughout the development of human society, the discovery and use of new materials play a critical and important role. Many stages in the human history are named after the popular materials used at the time. The first man-made stone tools began to appear in Africa about 2.5 million years ago, and metallurgical technology was not available until 4000–5000 years ago. Therefore, archeologists refer to the earliest prehistoric period that is longer than two million years as the Stone Age. That was the time period when early humans used stones as materials for making tools along with wood, bones, shells, antlers, and other natural materials. In the latter part of the so-called Stone Age, clay and other materials were sintered to make pottery, that short period is referred to as the pottery age. Later, with the advancement of time and with the development and innovation of a series of metallurgical technologies, mankind has experienced the Copper Age, Bronze Age, and Iron Age. Next, there was the Steam age, the Electrical age, the Atomic age which lead us to today, where we are advancing in the Information Age dominated by silicon materials.

There is no doubt that technology advances society and materials change how we conduct our daily lives. In addition, the implementation of materials is crucial to the safety of various application technologies. History of the development of human civilization is the account of how to improve and create new and better materials, and to use those materials reasonably and safely.

Some people may remember the explosion of the American Space Shuttle Challenger in 1986. This tragedy caused seven American astronauts including a civilian female teacher to lose their lives 72 s after takeoff. The famous physicist Richard Feynman participated in the investigation of the accident and demonstrated a simple "ice water experiment" to the public. It turned out that the physical reason behind the failure was a small O-ring washer. The earth-shattering accident was caused by a change in the nature of the material. Like an ordinary gasket, the purpose of the O-ring is to seal and prevent the hot gas of jet fuel from leaking out of the joint. However, due to the low temperature on the day of the space shuttle launch, the rubber material for the ring lost its elasticity, which made one of the O-rings ineffective, causing the hot gas to leak out and ignite the fuel in the external fuel tank, finally leading to a chain reaction of explosions.

In 1988, a near miss when a Boeing 737 passenger plane taking off from Hilo International Airport on Hawaii Island to Honolulu was damaged almost was causing a fatal air crash. The failure analysis later indicated a crack was formed from a small metal part that fatigue over time.

The importance of materials can also be seen in numerous Nobel Prizes in Physics and Chemistry awarded over 100 years. Among the Nobel Prize winners, those who have won prizes for discovering and researching materials account for a large proportion. These include the following examples.

German chemist Staudinger put forward the concept of macromolecules, which played a positive role in the vigorous development of plastics, synthetic rubber, synthetic fibers and other industries. Staudinger is the winner of the 1953 Nobel Prize in Chemistry for those discoveries. Nowadays, polymer synthetic materials, metallic materials, and inorganic non-metallic materials form three important categories in the world of materials.

Three American physicists, Shockley, Barding, and Bratton, shared the 1956 Nobel Prize in Physics for studying semiconductors and discovering the effect of transistors. Since then, the "silicon" era has started. The integrated circuits, which are indispensable in computer and information technology, has led the trend of science and technology for decades, are built on the basis of semiconductor silicon materials.

American Anderson, Van Fleck, and British Mott shared the 1977 Nobel Prize in Physics for their fundamental research on the electronic structure of magnetic and disordered systems. Anderson researched amorphous matter and created the localization theory of condensed matter physics, which enabled new superconducting materials to make great progress. Van Fleck and others made significant contributions to the study of antimagnetic and paramagnetic materials, while Mott studied transition solid materials such as metals.

In 1996, English chemist Harry Kroto and two American scientists Robert Curl and Richard Smalley won the Nobel Prize in Chemistry for discovering a new form of carbon-Fuller's sphere C-60.

The material we will introduce in this book—graphene, whose discoverer won the Nobel Prize in Physics in 2010, is another example of the connection between material research and the Nobel Prize. The Nobel laureates were two professors from the University of Manchester in the United Kingdom: Andre Geim and Konstantin Novoselov. Their groundbreaking experiment on two-dimensional material graphene has opened the door to the research and application of new nanomaterials. Graphene and more nanomaterials have received much attention in the most recent years. Graphene has many outstanding properties, such as high electrical conductivity, high specific surface area, high thermal conductivity and exceptional mechanical properties, and as a result has good application prospects in many fields.

Today's materials science is a multidisciplinary field involving the properties and applications of matter. With the rapid development of nanoscience and nanotechnology in recent years, materials science has been pushed to the forefront of technology.

What is nanotechnology? In terms of scale, nanotechnology refers to the study of materials properties and applications of materials with structure sizes ranging from

0.1 to 100 nanometers. Practically, the goal of this field of study is to directly use atoms or molecules to construct products with specific functions. As such, individual atoms and molecules are engineered to construct material structures.

The idea of nanotechnology came from Feynman. As early as 1959, Feynman predicted that we could start building structures from individual molecules, or even atoms, to assemble and control our requirements, which greatly expands the scope of our physical properties. This is the source of inspiration in developing nanotechnology.

In recent year, various high-potential materials are emerging, which one will be the main carrier in the next quantum era? For many, graphene is the front-running candidate.

Graphene is a two-dimensional nanomaterial (in terms of scale and structure). Although it can be said that small "fragments" of graphene are naturally present in graphite, the in-depth research and application cannot be achieved without nanotechnology. The discovery of graphene significantly promotes the development of nanomaterial synthesis technology. In addition, graphene involves profound physical theories and has substantial application prospects. The research and development of graphene can accelerate a wide application of new materials in various fields with principles involve quantum theory, special relativity, and topology in mathematics. Therefore, it can be said that graphene is a new material from the "ivory tower." Graphene is not only the new favorite materials for scientists, but also valuable in engineering applications. In theory, graphene can advance in-depth exploration of quantum theory with potential contribution to the next big breakthrough in theoretical physics.

However, the properties and theoretical studies of graphene mostly appear in professional books and journals. To the public, much of the science behind graphene is hard to understand; as a result, much of the basis and fundamental science is not available to the public.

So, what exactly is graphene? Why it has the magic? What are the related physical principles and application prospects? The market needs a popular science book that can explain this new material to the public in plain language and in simple terms, including graphene's physical nature. This is the purpose of this book.

We hope that this book can fill the gap between academic books and popular science. Reading this book will not only enable readers to increase their scientific knowledge, but also stimulate young people's interest in physics and materials science, leading them to the door of science and technology. In addition, the technical applications related to graphene not only require the updating of materials, but more importantly, the improvement of principles. This book will also benefit research and development personnel in various related fields, help to broaden ideas, be inspired, and carry forward this new material for the benefit of humankind.

As a historical review, in Chap. 1, the author shares stories of interesting and legendary scientific discovery of graphene. Chapter 2 is a brief introduction of quantum mechanics, which is necessary to understand graphene physics and to have a more vivid understanding of the microscopic world involved in graphene. Chapter 3 introduces several commonly used experimental detection methods used in graphene research to draw readers an intuitive image of the atomic and subatomic world.

The special properties of graphene come from its two-dimensional crystal structure. Graphene's electronic transport performance is closely related to energy band structure. A little bit of solid-state physics and energy band theory is explored in Chap. 4. Chapter 5 briefly explains the relativistic properties of the graphene energy band Dirac cone, and Chap. 6 introduces the relationship between graphene and topology from the perspective of popularity and interest.

The original meaning of the term graphene refers to a two-dimensional crystal with a single-layer atomic structure. The theory in the first half of this book and the interpretation of the wonderful properties and characteristics of this new material are aimed at graphene is an ideal atomic structure. However, "graphene" material prepared and applied in reality are far different from this ideal crystal. In Chaps. 7 and 8, the preparation methods of graphene are also briefly discussed, including introduction of various derivative materials and their application prospects.

To meet the reading interest of the public at all levels, the book utilizes simple illustrations combined with numerous examples and easy to understand terminology. The book can be served as a casual informative read or as an introduction literature for those who are just beginning their journey into the graphene world. The material covered inside is broad, including pure science such as quantum and topology, as well as materials related engineering technology. The author hopes that many fields in modern science and technology, including some basic theories in physics and mathematics, as well as a variety of applied technologies, can be advanced through the "magic graphene."

Chicago, USA   Tianrong Zhang

Acknowledgements I would like to thank Yi Sabia and Qiu Zhang for their contributions to make this English edition available. Yi Sabia helped revise some chapters and improve the text. Qiu Zhang helped to proofread the book.

Contents

Chapter 1
Discovery of Graphene

If you were to imagine the thinnest material in the world, how thick would you think it is? Your imagination will not be an unfounded guess, because you already have the basic knowledge of modern science. High school chemistry textbooks tell us: The smallest unit of a chemical element is an atom. So, most people can naturally conclude that the thinnest stable solid material should be made of a layer of atoms! Thinking of this, you may be a little excited, wow, a material made of a single atomic layer! What will its thickness be? What are its peculiar properties? How to make this material and apply it to technology?

To answer the above questions, we have to look at the basic atoms that make up this material. What is the composition? As it turns out, modern science and technology have provided us with an excellent example. This thinnest material has been manufactured. The research and development of graphene has a history of more than ten years, and it is beginning to be recognized as material of choice for many applications. This material, graphene, is slowly penetrating our daily lives and will be introduced to you in this chapter.

1.1 What Is Graphene?

Graphene is a two-dimensional honeycomb network lattice structure composed of a single layer of carbon atoms closely packed, which looks like a plane (or approximately a plane) composed of a hexagonal grid. Each carbon atom forms a regular hexagon with the surrounding carbon atoms through three covalent bonds, as shown in Fig. 1.1. The thickness of single-layer graphene is only 0.335 nm (nanometers), which are about one-200,000th of the diameter of a human hair. It is the thinnest material in the world.

The name "graphene" is derived from the mineral graphite, which is the material used in pencil refills that everyone is familiar with. The traditional name "lead pencils" is a historical misunderstanding. In fact, there is no lead in pencils; instead

© Guangxi Science & Technology Publishing House 2022
T. Zhang, *Graphene*, https://doi.org/10.1007/978-981-16-4589-1_1

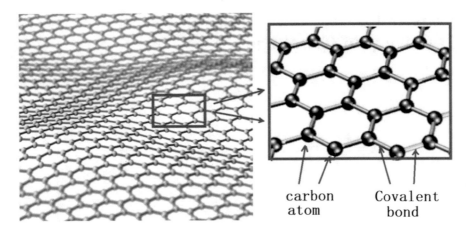

carbon   Covalent
atom   bond

Fig. 1.1 Hexagonal network structure of graphene

their component is graphite, which is made of carbon atoms. As early as the sixteenth century AD, British discovered a large amount of black minerals in a place called Borrowdale. This mineral is dark and shiny, and the local shepherds often use it to mark their sheep. Several people who observed this mineral were inspired by this application and thought: if this mineral can be used to mark sheepskin, it should be able to leave marks on paper, and thus can be used for writing! However, they mistakenly thought that this was the same material as the lead that the ancient Romans wrote with on paper. Even though this material was softer and darker than lead and the writing was much clearer and more beautiful. As such they called this black mineral "Black lead".

Soon after, King George II collected the current day Borrowdale Graphite Mine as the royal family's possession and designated it as a royal patent. In 1761, the German chemist Kaspar Faber turned graphite into graphite powder, mixed with sulfur and other substances to form a finished product, and then sandwiched them in a wooden strip to become the earliest pencil. Since then, the pencil industry has prospered with the mining of the Borrowdale graphite mine. Now, more than 400 years have passed. If you travel to Borrowdale, you can see the museum in nearby Keswick. The largest pencil in the world is on display, recording the traces of this period of history.

Swedish chemist Carl Scheele discovered that the mineral named "black lead" was not actually lead in 1778, but rather composed of carbon atoms. Later, German geologist Abraham Gottlob Werner changed the name of this substance from "black lead" to Graphite, because the word means "writing" in Greek.

The graphite discovered by the British in Borrowdale has filled the wallets of many merchants and introduced the pencil to the world, in turn enabled the recording and spreading of human culture. However, they may never have imagined that today's scientists have created an ultra-thin, ultra-strong and ultra-transparent material from that dark and soft graphite. This is graphene. Where does "ene" come from? It is derived from the description of single atomic layer structure in chemistry.

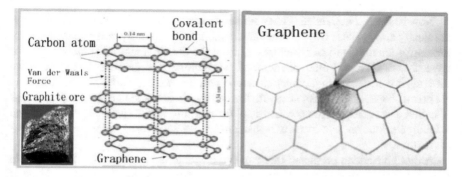

Fig. 1.2 From graphite to graphene

Now that we have a preliminary understanding of graphene, we look back at the structure of graphite and find that graphite is actually composed of discrete layers of graphene stacked on top of each other, just like a deck of playing cards. In other words, graphene is the thinnest layer in the graphite structure. When we swipe a pencil on the paper, maybe a small piece of graphene is created! See Fig. 1.2.

Although graphene and graphite sound similar they are actually very different from each other. Graphite is the raw material mined from the ground while graphene is a single layer of graphite discovered in the lab.

In fact, in the early days, theoretical physicists were not optimistic about the likelihood of single-atom layers of 2-dimensional materials, as they thought they were too unstable to exist. For example, a famous theoretical physicist in the former Soviet Union, Lev Landau (1908–1968), theoretically proved the instability of 2-dimensional crystals in the 1930s [1].

Imagine: trying to create a two-dimensional material with a single layer of atoms, the thickness of which is only one-200,000th of a human hair. This sounds a bit like an impossible fantasy! More than 80 years ago, even a master of physicist like Landau could hardly predict that humans could actually create such materials. However, even though it is not achievable in the laboratory, it can always be discussed in theory. This is the charm of science.

Landau et al. believed that from the perspective of thermodynamics and statistical physics, when the absolute temperature T is close to 0, two-dimensional lattices of any size may exist. However, as T increases, the vibration energy of lattice system also increases. Some energy will go out of the plane, causing certain atoms to fly away from the two-dimensional structure. In other words, under the condition of nonzero temperature, the thermal fluctuation of the two-dimensional crystal will destroy its structure, making it extremely unstable. Therefore, Landau et al. concluded that two-dimensional materials cannot exist in nature at room temperature.

Landau is an expert in the study of crystals, and the founder of solid state and condensed matter physics. His views and judgments are not trivial, which makes most people stay away from the experimental development of two-dimensional materials.

Now it seems that Landau is not completely wrong. In order to obtain a thermody-namically stable state, two-dimensional crystals will naturally curl and easily form tubular or small spherical structures. During the development of graphene, people have also observed similar phenomena. The thinner the graphite layer, the more difficult it is to maintain a planar structure, and it is easy to curl into a columnar or spherical shape. A columnar structure formed by a hexagonal lattice of carbon atoms is called a carbon nanotube, and if it is formed into a spherical shape, it is called a fullerene. Materials of these two structures were discovered before graphene was discovered.

Although Landau predicted that two-dimensional lattices are difficult to exist in isolation, there are always attempts to create two-dimensional materials. Even if they are unstable, you can find a way to explore and study some of the new physical properties. Moreover, as far as the graphene monolayer atomic two-dimensional crystals obtained today are concerned, they are all attached to a certain "substrate". They do not need to float completely alone in the air.

1.2   Andre Geim's Random Walk

Since graphite is slices of graphene stacked together, it means that graphene originally exists in nature, in graphite ore, that is, in the pencil lead that people often use. However, it is extremely difficult to draw one layer from this pile. So, how do you get 2-dimensional graphene from the 3-dimensional graphite, layer by layer? Obviously this has been demonstrated, the process itself was not easy and went through a lot of difficulties with twists and turns, and there are many interesting stories interspersed with them.

In the early 1990s, people began to study 0-dimensional carbon nanospheres and 1-dimensional carbon nanotubes, but they had not yet tackled 2-dimensional carbon structures. The resources of graphite are extremely abundant and common on the earth, and humans' high technology has reached the point where we can clearly see the atoms, and under certain conditions, we can also manipulate atom one by one. How can we find a way to separate discrete thin layers of carbon atoms from graphite, or even a single layer of "graphene"?

In 1990, a German physicist used the method of scraping graphite on the surface of another substance to create thin and transparent graphite sheets. He named this method "micromechanical cleavage". However, although this cleaved material has been "thinned to transparency", it was far from a single layer of atoms.

Then, Philip Kim, a South Korean professor in the Department of Physics at Columbia University, was interested in single-layer two-dimensional crystals of carbon atoms and attempted to use a similar micromechanical splitting method to separate the thin graphite layer. In 2002, Philip Kim guided one of his Chinese PhD students to start researching this subject. This Chinese student tried to obtain graphene by writing with a pencil. He spent two years to research and manufacture a very small "nano pencil", and using it to get about 30 layers of carbon.

Just as Philip Kim and others were excited about the results of their "30 layers of carbon atoms", physicists Andre Geim and Konstantin Novosolo from the University of Manchester in the United Kingdom published their article in the "Science" magazine [2], announcing that y had successfully made a single-layer graphene!

How did they succeed? It turns out that the University of Manchester research team had been trying to separate graphene from graphite since 2000. The lead professor Geim is not one who would imagine what a physicist is. Before we tell the story of how he used sticky tape to win the Nobel Prize, let us first review the stories of his other scientific research achievements, which will definitely transform the stereotyped image of "physicists".

Andre Geim (1958–) was born in Sochi, Russia in 1958, a small city on the Black Sea, where Geim's parents were both engineers. Geim went to the Moscow Institute of Physics to receive his higher education, and later received a doctorate from the Institute of Solid State Physics of the Russian Academy of Sciences. After graduating and working at the school for three years, he continued his research in the UK, Europe, Denmark, and the Netherlands. Geim has Dutch nationality and is currently employed by the University of Manchester in the UK.

Geim won the 2010 Nobel Prize in Physics together with his student Konstantin Novoselov (1974–) for "a pioneering experiment in two-dimensional graphene materials". Geim gave a fantastic Nobel speech on this topic, named "A Random Walk to Graphene" [3], see Fig. 1.3. The scientific method and innovative thinking explored in Geim's speech were refreshing and mind-boggling; his humorous language and intriguing examples won endless laughter and applause.

Geim's "random walk" refers not only to the constantly changing "walk" in the geographical location as shown in Fig. 1.3, but also to a deeper meaning of his various scientific research experimental subjects. A complex and interesting "walk" guided by lateral referencing. Geim's scientific research walk has confirmed the importance of cross boundary research.

Geim is a physicist, but his first famous research work is related to frogs. He and Michael Berry, a physicist who is famous for proposing geometric phases, studied

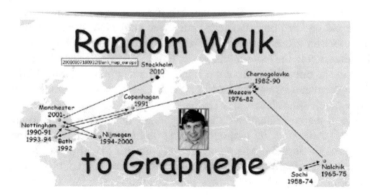

Fig. 1.3 The "Scientific Walk" of Geim, the Father of Graphene (from Geim's Nobel Lecture)

"The Physics of Flying Frogs" together, and won the 2000 Ig Nobel Prize for Physics [4].

Most people know the Nobel Prize, but they don't necessarily know the "Ig Nobel Prize"; many people have heard of the "Maglev train", but not necessarily the "Magnetic Levitation Frog". Geim won the Ig Nobel Prize in 2000 for his research on magnetic levitation of frogs.

The Funny "Ig Nobel Prize" is famous for grotesque but scientifically meaningful research. In fact, it is not just a joke and ridicule, but more of a kind of humor in academia. It is said that its purpose is: "First, let you laugh. Then let you think." There are many creative people among the Ig Nobel winners. For example, Geim can definitely be counted as one.

Professor Geim is the first and only double winner have won both the Ig Nobel Prize and the real Nobel Prize. So, what is going on with this magnetic levitation frog?

1.3 Flying Frog and Gecko Tape

In the 1990s, after Geim received his doctorate from Russia, he once worked as an associate professor at the University of Nijmegen in the Netherlands. At that time, the main equipment advantage of his laboratory was the powerful electromagnet. These devices can generate a magnetic field of about 20T, but unfortunately, Geim's research topic at that time only required a weak magnetic field of less than 0.01T. This inspired Geim's "lateral thinking", as he was always trying to find a subject that could use such a powerful electromagnetic field.

Geim was inspired by the phenomenon of "magnetized water". It was not the bottled magnetized water sold commonly in supermarkets; rather the idea of putting a small magnet around a hot water pipe prevents formation of scale inside the pipe. It sounds reasonable. Water is not a ferromagnetic substance, but media like water generally have the so-called "diamagnetic". That is to say, under normal circumstances, the electronic magnetic moments of atoms in diamagnetic materials cancel each other out and are non-magnetic, but when subjected to an external magnetic field, the electron orbital motion will change, causing a magnetic moment in the opposite direction to the external magnetic field. However, this diamagnetic effect is extremely weak, one billion times weaker than the magnetism of iron! The magnetic field of the permanent magnet placed on the faucet is also very small, and it is hard to say how much effect it can play on the scale.

Looking at the equipment with a strong magnetic field in the laboratory, Geim was inspired: Although the magnetic susceptibility of water is very small, under this strong magnetic field, water may be magnetized! So, how will it behave after magnetization? Curiosity drove Geim to make an unusual and deviant operation. On a Friday night, he stupidly and slowly poured a little water into the instrument that was generating a huge magnetic field...

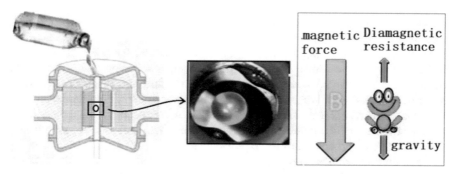

Fig. 1.4   Water droplet floating in a strong magnetic field and its principle

   The result was surprising. The water did not flow out of the strong magnet, but gathered into a water ball about 5 cm in diameter, which floated freely in the center of the magnet, as shown in Fig. 1.4: gravity disappeared, like the object is floating in space! With such a weak reverse magnetic effect of water, the negative magnetic susceptibility is generally about 10^{-5} (in SI unit), but it can actually resist gravity. Geim was excited, and like a naughty child, he continued to "throw" things into the magnetic field: strawberries, tomatoes, insects… and even frogs!

   It is this "flying frog" that can only resist gravity and is suspended in a magnetic field that made Geim and another famous theoretical physicist Berry (Sir Michael Berry, 1941–) win the 2000 Ig Nobel Physics prize. At the beginning, the awarding party of the Ig Nobel Prize asked Geim and Berry for their opinions: Do you have the guts to accept the award? They readily agreed, which fully demonstrated the sense of humor and self-deprecating courage of the two physicists.

   Although this research was awarded the "funny" Nobel Prize, the results of the experiment contained profound meaning, and it showed diamagnetism intuitively. It reminds people: "Non-magnetic" substances, including biological and human magnetism, are not negligible, and under certain conditions, they can be large enough to resist gravity. It also makes people realize that it is possible to carry out certain biological experiments under similar space conditions in a laboratory with a sufficiently large magnetic field.

   For Geim, this research has benefited him a lot. First, his popularity increased greatly. The newspapers and media rushed to report that year. It is said that until now, many young scholars said hello to Geim: "Professor, I have known you a long time! But not because of graphene, rather because of floating frogs."

   From this interesting experiment that made frogs fly, Geim realized the importance of lateral thinking for scientific research, especially when helping young students choose topics, it is very important to stimulate their pursuit of fun and let them integrate scientific research into entertainment. Sometimes, trying to do some research that seems inconsistent with your own professionalism may produce very important and interesting results. Since then, Geim has started to do some unconventional experiments and calls them "Friday Night Experiments".

Such attempts by Geim and his team often end in failure, but the probability of success is not small, such as his second experiment "gecko tape".

Geckos are common reptiles with super climbing ability. They can climb on any surface, even glass ceilings. The answer to this super adhesion ability is that the gecko's toes are covered with a very fine hair. Each fluff and surface can produce extremely weak Van der Waals forces, but the combined effect of countless such fluffs cannot be underestimated. These are the results of research by biologists. Geim happened to see their article, and his brain flashed: Why not develop an artificial adhesive material based on this principle?

Geim and his friends really designed this material and named it "gecko tape". Although the adhesion performance of this artificial material after repeated use is not as strong as the toe of the gecko, it is similar. The method of thinking has greatly inspired physicists and materials scientists to study bionic materials.

Professor Geim has another incredible thing: In 2001, he co-authored a paper with the author named "H. A. M. S. ter Tisha". Can you imagine the dignity of Geim as a co-author of the article? It turns out that this is his pet, the name of a hamster, see Fig. 1.5.

Why did the hamster become Geim's collaborator? It turns out that this article is an application of the "magnetic levitation frog" experiment. Geim uses the diamagnetism of objects to conduct an "antimagnetic levitation gyroscope" experiment to detect the rotation of the earth. He believes that his hamster was indispensable in the process of his experiment. As an important suspended substance in the experiment, this hamster is qualified to be a co-author of the paper.

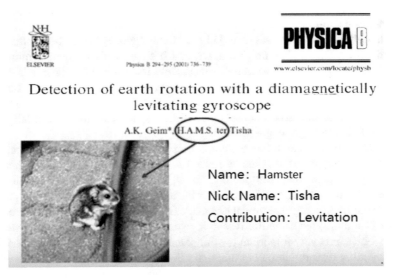

Fig. 1.5 The paper by Geim and the hamster [5]

1.4 Sticky Tape Led to Nobel Prize

From the two research examples in the previous section, you have a general under-standing of Geim's scientific research style. He is a person who finds fun in scientific research. Now, let's go back to his discovery of graphene.

As Professor Geim began to imagine a two-dimensional crystal material of carbon atoms, a new Chinese student Da joined his team. Newcomers are expected to be familiar with the laboratory environment, to work to improve their English, and hone their experimental skills. Therefore, Geim assigned Da to polish the graphite sample with an advanced polishing machine, which can grind the sample to a flatness of a few tenths of a micrometer.

Of course, Professor Geim did not inexplicably give the students the work of "grinding graphite rods". This idea came from several questions that Geim had pondered for months. In his Nobel lecture, Geim summed up these issues into the "three little clouds" [5].

The first cloud is about metallic electronics. Everyone knows that metals have good electrical conductivity, and semiconductors are just between insulators and metals. However, why are semiconductors widely used as materials for the manu-facture of integrated circuits and devices, while metals are not? Experts will tell you: Because the performance of semiconductor materials will change with the addition of different impurities, n-type semiconductors and p-type semiconductors are formed, from which p-n junctions are manufactured and applied as various transistors. This is indeed the theoretical basis for the work of modern semiconductor electronic devices, but is this kind of historical development the only way to achieve modern scientific and technological civilization? For example, the surface conductivity of metals can also be changed under different conditions. Such changes are extremely small, but it is possible to have applications. In particular, if a metal or semi-metal is made into a thin film, its electric field effect maybe possible to use for some special applications.

The second little cloud is about carbon nanotubes, a popular research in the 1990s. Geim was attracted by the beautiful research results displayed in this field and wanted to engage such research. However, he feels that the most prosperous time has passed.

The two small clouds staying in Geim's brain sparked brilliant thoughts due to the appearance of the third small cloud. That was the inspiration Geim got after reading an article about inserting other substances between layers of graphite. It is mentioned that although scientists have studied graphite for many years, they still know very little about the material. Connecting this little cloud with the previous two: the excellent properties of graphite, thin films, carbon nanotubes... Geim felt that graphite was worthy of further study.

So Geim conceived a topic from three small clouds, and let his new graduate student Da try it first.

Da grinds for three full weeks, and grinds out a slice of 10 μ thick graphite, about the thickness of 1000 carbon atoms. Obviously, this result was still "a long way from the monoatomic layer."

This is a huge setback, is it worth continuing? There are many projects, and there is more than one path to scientific research. Just as Geim and others thought about it, they didn't expect an unrelated thing to change their minds.

Working next to Geim's laboratory was Oleg Shklyarevskii, a senior fellow from Ukraine and an expert in scanning tunneling microscopy (STM). Geim discussed with Shklyarevskii about his graphite polishing work, jokingly saying that it was like "polishing a mountain to get one grain of sand"! Unexpectedly, after listening to this, Oleg fished out some adhesive tape from a litter bin, and gave it to Geim. The STM expert explained to Geim that graphite is a common sample type. Before experimenting on graphite, the sample being placed under the piezoelectric tip must is clean. The method to ensure graphite cleanliness, adopted by STM field, is simple and quick. Adhere Scotch tape to the top of the sample will remove a layer of the graphite, along with any surface contaminates. Oleg stated that no one had ever carefully looked at what was on the discarded tape. As a result of the discussion, Geim inspected these Scotch tape generated graphite samples microscopically, and observed that some fragments were much thinner than the ones polished with the machine. At this time, Geim realized that how silly it was to suggest Da using polishing machine. Why not use tape?

Unfortunately, Oleg was busy with his own experiments and did not participate in Geim's "tape stripping method", but Konstantin Novoselov participated. Novoselov was a young researcher, under 30 years old. He was a student of Geim at the University of Nijmegen in the Netherlands. He and Professor Geim hit it off because of their common Russian background. Novoselov also received higher education at the Moscow Institute of Physics and Technology in Russia. The valuable thing is that Novoselov had studied engineering, had excellent experimental skills, and was very interested in the "Friday Night Experiment". He has already participated in Geim's research on gecko tape.

So Geim and Novoselov began to use scotch tape to deal with graphite: sticking, tearing, sticking, and tearing, repeated experiments many times, finally they generate a very thin sheet. Then, Novoselov thought of using tweezers to move the stripped graphite flakes from the tape to a silicon oxide wafer substrate, so that the corresponding measurement could be made by applying voltage to the silicon. The measurement results showed that some of the graphite flakes were only a few nanometers thick, which made the two of them excited. In this way, the first two-dimensional crystal material, graphene, officially appeared on the stage!

In October 2004, the "Science" magazine published the paper by Geim and Novoselov. In 2010, the two scholars won the Nobel Prize in Physics for their results.

1.5 Carbon Atom Family

As explained, graphene is just a single layer of carbon atoms. Carbon is a very common in our lives: carbon dioxide, carbon monoxide, coal… Carbon atoms are everywhere. Its abundance in the universe is only less than hydrogen, helium, and

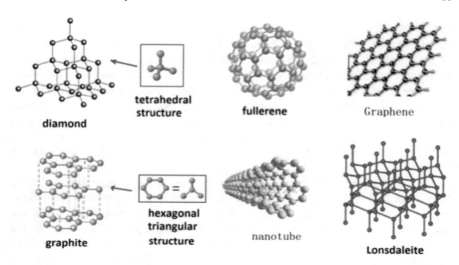

Fig. 1.6 Allotropes of carbon (https://courses.lumenlearning.com/introchem/chapter/allotropes-of-carbon/)

oxygen. It is extremely widespread on the surface of the earth and is the main element of organisms. The carbon cycle on earth is closely related to the evolution and development of life.

In addition to various compounds of carbon, carbon has a variety of allotropes, that is, various molecular structures, forming a large family of carbon, as shown in Fig. 1.6. Among them, by the end of 2017, graphene is still the youngest member of the big family.

Although the various allotropes above are composed of carbon atoms, their physical properties are completely different.

For example, take the oldest member diamond (Fig. 1.6, upper left) and graphite (Fig. 1.6, lower left). Many of their physical properties are located at opposite extremes. Graphite is soft, whereas diamond is the hardest ore. Graphite is a good conductor and diamond is an insulator. Graphite is jet black and opaque, and diamonds are crystal clear and shiny. Graphite is very common and everywhere, but diamonds are rare to find.

The huge differences in the physical properties of graphite and diamond come from the difference in the arrangement of carbon atoms. As mentioned earlier, graphite is composed of two-dimensional lattice flakes, which overlap like playing cards randomly. All the carbon atoms in diamond hold hands to form a firm three-dimensional crystal structure.

Diamonds are extremely rare, and the exact age of their earliest discovery is difficult to verify. In India and China, there are similar records of wearing diamonds and cutting objects with its super hardness in BC. However, it is recognized in academic circles that the first primary diamond ore was discovered in South Africa in 1871, and a 3,106 carat diamond was discovered in the "Prime" Kimberley Cave in South Africa in 1905, which is the largest gem-quality diamond in the world today.

By 1779 and 1794, scientists were able to determine the composition and structure of graphite and diamond respectively. Soft dark graphite and hard shiny diamonds are made of carbon atoms. Under different conditions, carbon atoms are arranged in different ways to form different substances. Under higher pressure and temperature, carbon atoms crystallize to form a hexagonal two-dimensional structure, and then overlap to form graphite; under extremely high temperature and pressure, carbon atoms are arranged into a tetrahedral three-dimensional crystal structure, which is diamond.

Most people believe that natural diamonds are formed due to high temperature and high pressure deep in the earth, and then are brought from deep underground to the surface by topographic changes such as volcanic eruptions. This method of formation makes natural diamonds rare and expensive. As such, princes and nobles of past dynasties valued diamonds as a symbol of wealth, while merchants fought to the death for natural diamonds. Now, scientists already know what the structure of diamonds is all about. Its structure seems simple and beautiful, which triggers people's desire for artificial diamonds. Many people consider creating artificial high temperature and high pressure conditions in the laboratory to synthesize artificial diamonds. However, the idea of artificial synthesis is extremely difficult to implement. Finally, General Electric announced its first success in 1955, which opened up a promising market for diamond applications. It has to mention the contribution from Howard Tracy Hall (1919–2008).

In order to synthesize synthetic diamonds, it is necessary to create a high-temperature, high-pressure, oxygen-deficient environment in the laboratory. For this purpose, General Electric planned to purchase a large press worth $125,000. At that time, Hall was just a small employee of the company, but he proposed a plan that could work for only about a thousand dollars. However, Hall's plan was rejected by the company's management as a whimsical joke. Hall pursued his concept without managerial support, using an old pressure pump to explore for several years, and continuously improved to produce an artificial environment higher than 10 GPa and a temperature higher than 2000 °C, and finally succeeded on December 16, 1954 to synthesize artificial diamonds.

In addition to graphite and diamond, carbon allotropes include activated carbon, carbon black, coal, and carbon fiber and other amorphous forms. However, this book is only interested in crystalline carbon materials (Fig. 1.6).

The reason for the different physical properties of various allotropes of carbon is that the microstructures of their carbon atoms are different, and the different crystal arrangements result in different "hybridization modes" between the outer electron orbits of carbon atoms. The concept of "orbital hybridization" is the application of quantum theory in the theory of chemical valence bonds. In the next chapter, quantum mechanics will be specifically introduced. In the following description, readers can first understand orbital hybridization from the perspective of classical chemical bonds.

In classical mechanics, the motion of electrons in molecules or crystals has been described using orbital model, in which electrons orbit nucleus like a planet. But in quantum mechanics, electrons can no longer be regarded as solid particles with a fixed shape. The so-called "orbital", that is, the wave function solved from the quantum mechanics equation, is completely different from the classic "orbit". Orbital lines are complex functions permeating the entire space. However, in the generally recognized Bohr semi-classical atomic model, the term "atomic orbital" is still used. It is described by the "probability wave" of electrons represented by the wave function, that is, the concept of "electron cloud" similar to the atmosphere is used instead of the classic linear orbit.

According to the Bohr model, when the number of electrons in an atom is more than one, the wave functions are superimposed on each other, and the actual shape of electron cloud is not easily decomposed into ideal images of individual orbits to describe. However, from the mathematical point of view, wave functions can still be simplified into a combination of simpler atomic orbital functions. Therefore, in a certain sense, the electron cloud can still be imagined as a superposition of atomic orbitals of different shapes, and each orbital contains one or two electrons. That is, the atomic model in quantum mechanics is still often explained by intuitive classical orbital images.

In other words, the atomic orbit of the Bohr model is much different from the elliptical linear orbit of the planet. Bohr-orbit is a series of wave functions composed of three variables (quantum numbers n, l, m), which can be intuitively imagined as the electronic cloud shown in Fig. 1.7.

The quantum number n describes the energy degenerate orbit according to different energy values, which is represented by the first number 1, 2, 3, etc. in Fig. 1.7. Then, according to the angular quantum number l, it is divided into s, p, d orbitals. This classification gives an intuitive description of the orbital electron cloud: the s orbital is spherically symmetric, the p orbital is axisymmetric, and the d orbital extends in 4 directions…, as shown in Fig. 1.7.

There are 6 electrons outside the carbon nucleus and 4 valence electrons in the outermost layer. Two or more atoms in the crystal use their outer electrons together to form a covalent bond under ideal conditions, and the composition is stable when

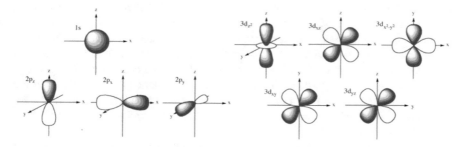

Fig. 1.7 The electron cloud describes the s, p, and d orbitals (https://www.pngwing.com/en/search?q=orbital+Overlap)

the electrons are saturated. The structure of the outer carbon atom contains 4 electrons including one 2s orbital and three 2p orbitals. However, generally speaking, in a crystal composed of carbon atoms, the covalent bonds of atoms are obviously not formed by these one 2s and three 2p orbitals. The redeployment of these 4 orbitals produces four new orbitals. This process is called heterogeneous. The new track is called hybrid track. Different hybridization methods produce different hybrid orbitals.

Simply put, the hybridization method represents the shape of the electron cloud formed by the electron orbits near the atoms, which is used to describe the linking method of covalent bonds between atoms.

There are 3 different ways of hybridization between carbon atoms (sp, sp^2, sp^3), which can form crystal structures with different physical and chemical properties. Carbon is a four-valent element, and each carbon atom has four valence electrons in the outermost layer. In the crystal structure of diamond, the four valence electrons of each carbon atom are involved in the formation of covalent bonds, as shown in Fig. 1.6a. The tetrahedral structure of diamond corresponds to the hybrid orbital sp^3. All the electrons in it form covalent bonds and there are no free electrons, so diamond does not conduct electricity. Each atom is tightly bound to the other 4 carbon atoms and cannot be separated easily, so the diamond is very hard.

Graphite is different. In the layered structure, each carbon atom forms 3 covalent bonds with the other 3 carbon atoms, forming a flat hexagonal (or triangular) tiled structure, as shown in the lower left corner of Fig. 1.6, it corresponds to the hybrid orbital sp^2. Therefore, each carbon atom in the graphite layer has one electron left. They dissociate between the layers and are shared by the atoms in the upper and lower layers. Together, they form a π-cloud throughout the entire layer and become free electrons that can flow. Therefore, graphite has excellent electrical conductivity and thermal conductivity. From a chemical point of view, graphite is equivalent to a weak van der Waals force between each layer. The layers can slide in parallel easily, so graphite is soft and can be used as a pencil lead to write on papers or used as a lubricant.

Some other allotropic structures of carbon, such as carbon nanotubes, graphene, and fullerenes, are basically dominated by plane regular hexagonal structures similar to those in the graphite layer, sometimes mixed with pentagons and heptagons. If you ignore the small nanoscale and classify it from the geometric dimension, carbon nanotubes can be regarded as a one-dimensional structure, because their scale in the radial direction is very small, on the order of nanometers, and tens of thousands of carbon nanotubes Together, it is only the thickness of a hair strand, and it is much larger in the length direction, up to tens to hundreds of microns. Graphene is a two-dimensional structure. Each carbon atom in the plane and the other 3 carbon atoms form a strong covalent bond, so graphene is the strongest and thinnest two-dimensional material. Theoretically, at about the same thickness, the strength of graphene can reach 200 times that of steel.

Fullerenes can be regarded as point-like 0-dimensional structures. They present a variety of shapes, but all are approximately spherical with a nanometer radius. In fact, many of them are polyhedral shapes, as shown in Fig. 1.8. The most typical

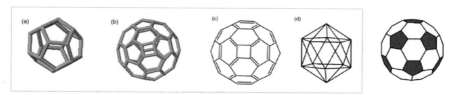

Fig. 1.8 Fullerene

fullerene is the buck ball C60 with 60 carbon atoms, and its structure is similar to a modern football.

Shown in the lower right corner of Fig. 1.6 is another allotrope of carbon. It has a wonderful name: Lonsdaleite, which is a kind of hexagonal diamond. The first discovery of naturally-occurring Lonsdaleite stone was found in a meteorite. Scientists believe that Lonsdaleite was formed when graphite in the meteor was introduced to huge pressure and heat is the meteor fell to earth. Its crystal structure is different from that of diamond, as it has a hexagonal lattice structure. The strength of Lonsdaleite in meteorites is slightly inferior to that of diamonds, but the strength of Lonsdaleite produced by artificial synthesis is about 58% higher than that of diamonds. People explain that it is because of the natural formation of Lonsdaleite includes structural defects from impurities.

Each member of the carbon allotrope family has its own unique characteristics, especially the emergence of graphene series in recent years. In particular, graphene has very good electron transport properties, and is known as the "king of new materials." Some people predict that it may play a disruptive role in the electronics industry, and potentially may even save the impending failure of Moore's Law for the semiconductor industry In terms of micro-structure; graphene (the hexagon) is the basic unit that constitutes the structure of fullerene, carbon nanotube, and graphite. Understanding the electrical properties of graphene helps to understand the physical properties of similar structures. Therefore, we will introduce the physical basis of graphene's electronic properties in more depth.

The discovery of graphene seems to be the result of repeated pasting of adhesive tape, but in fact, graphene, a new nano-level material, has a nearly perfect atomic structure. The quantum tunneling effect obtained is a peculiar phenomenon caused by the coupling of spin and orbit. Many peculiar properties of graphene are difficult to explain with classical physical theories. Therefore, we first introduce a little quantum mechanics in the next chapter, and then start with the basic knowledge of quantum theory and energy band theory in solids.

Chapter 2
The Quantum Mechanics

2.1 Quantum Essences

In 1900, German physicist Max Planck (1858–1947) proposed quantum concept to solve the problem of blackbody radiation. Even though at the present quantum theory already has a history of more than 100 years, experts are still debating on how to interpret quantum mechanics. For the general public, the theory is even more uncertain and baffling. However, on the other hand, quantum mechanics can be described as a model of success theory. The technology is widely used in semiconductor, materials science, chemistry, biology, and other fields. It is no overstatement to say that without quantum theory, there would be no modern society with such advanced technology. Quantum mechanics has penetrated into all aspects of modern materials science and quantum theory can be used to understand the physics of enchanting graphene. On the contrary, the in-depth study of the structure and properties of graphene can enhance our knowledge of quantum theory.

We introduced the physicist Landau's statement about 2D crystals in the first section of Chap. 1. He believed that 2D crystals are unstable and may not exist alone. Landau's conclusions were drawn from his research on the application of quantum mechanics to crystals. Now that we have produced 2-dimensional crystalline graphene, we need to use quantum theory to explain the principle that it can exist independently under certain conditions in order to understand the physical properties of this material. The purpose of this chapter is to let readers understand the basic concepts of quantum theory.

Quantum theory is a set of physical laws applicable to microscopic scales (such as atoms and electrons). Graphene is a thin film composed of a single layer of atoms, with a thickness of about one-third of a nanometer, and only one-200,000th of the diameter of a hair. Under such small scale, physics follows quantum laws.

At the atomic scale, the classical mechanics has failed. Quantum phenomena are quite different from those that can be explained by the classical Newtonian theory in our daily lives. Physicists established Newton's classical mechanics and classical electromagnetic theory on the macroscopic basis. However, the microscopic

© Guangxi Science & Technology Publishing House 2022
T. Zhang, *Graphene*, https://doi.org/10.1007/978-981-16-4589-1_2

world described by quantum mechanics can be said to completely lose the direct observability of human senses. For example, you can feel electric current, but you cannot "directly" perceive an electron or proton. You can see various colors with your eyes, but you can't see the photons one by one. For another example, the graphene structure cannot be seen directly with the naked eye. As for concepts such as quark, the ability to use our senses diminishes further. In other words, the small size of the microscopic world makes it impossible for humans to experience it intuitively, and can only be measured indirectly using certain experimental methods, and described imaginatively using abstract mathematical models.

Because the microscopic world is difficult to observe with the naked eye, when it comes to quantum theory, many readers may only think of the strange and unreasonable phenomena introduced in many popular science books, such as "Schrödinger's Cat", etc.

To sum up, there are three main confusions about quantum theory: one is at the physics level, which comes from the essential difference between microscopic and macroscopic phenomena; the other is from the interpretation, which comes from different physicists. In this book the Copenhagen interpretation is used for most of time; the third one is from public misunderstanding of quantum phenomena. As mentioned above, only the physics is essential and unavoidable. The other two confusions are avoided in this book.

Why are the laws of quantum physics different from the classical laws? The root cause is simple. It is just that certain physical quantities describing the motion of microscopic particles need to be "quantized" when transitioning from the macro to the micro.

The term "quantization" here is different from the term "secondary quantization" used in quantum field theory in physics. What we call quantization refers to the transition from classical to quantum; some physical quantities will change from continuous to discrete. In other words, the term quantum represents a certain discontinuity. Sometimes, certain physical quantities can only take discrete values. This is not difficult to understand. For example, from a human macroscopic scale, a section of slope formed by sand is quite smooth, but to a tiny ant, it looks like a staircase with one step and another step. Of course, the meaning of quantization is far from the above metaphor, but it can help beginners to understand. In other words, in classical physics, there seems to be no limit to the minimum change of physical quantities. They can change arbitrarily and continuously, and theoretically they can be as small as possible. But in quantum mechanics, because the object to be processed is the microscopic world, the spatial location of the physical object and other physical quantities can generally only be changed one by one in a certain value.

The concept of energy quantization was first proposed by Planck in 1900 [6]. This is not out of his inexplicable conjecture, but to solve a problem of "black body radiation" that the experiment does not conform to the classical theory.

The "black body" here does not necessarily have to be "black", it refers to an ideal object that only absorbs but does not reflect or refract. But, the "black body" still has radiation. For example, a dark fire stick is not always "black". When it is placed in an iron furnace, its color will change with temperature changes: first, after the temperature gradually rises, it will turn dark red, then brighter red, then bright golden yellow, and later, it may appear blue and white. Why are there different colors? This shows that at different temperatures, the poker radiates light of different wavelengths, which is black body radiation.

But in Planck's time, the classical theory of blackbody radiation ran into difficulties, which was far from the experimental results. Planck solved this problem and got results that were in good agreement with the experiment. He adopted an ingenious and novel method of thinking, which is to assume that when a black body radiates, the energy is not continuous, but is emitted one by one with a certain minimum value. In other words, the concept of "energy quantization" is introduced. In order to limit the minimum value of radiant energy, Planck assumed a Planck constant h. For more than 100 years, the appearance of this constant has become a sign of quantum world. Five years later, Einstein (1879–1955) further proposed the concept of light quanta and successfully explained another physical phenomenon that cannot be explained by classical theory: the photoelectric effect [7].

In 1912, Danish physicist Niels Bohr (1885–1962) used the concept of quantization to establish a new atom model [8]. He believed that atoms can only stably exist in a series of discrete energy states. In order to separate the stationary state, any energy change in the atom can only be carried out in a transition between two stationary states. Therefore, the electrons in the atom can only be in a series of discrete stationary states.

Black body radiation, photoelectric effect, and Bohr Atomic Model, these closely related concepts, along with Planck's constant h, formed the backbone of the quantum concept.

2.1.1 Wave-Particle Duality

In classical physics, particles and waves are two completely different physical phenomena, but in quantum theory, wave-particle duality is the basic property of all microscopic particles. Whether it is atoms, electrons, or light, they are both particles and waves..

Basic physical constants often play an important role in theory, and Planck's constant h in quantum mechanics is a good example. Planck uses it to quantize electromagnetic wave energy. Electromagnetic wave is continuous in the classical sense, but energy is quantized in the microscopic world. The smallest possible units (photon) by which energy of electromagnetic wave can be transferred is $h\nu$, which is equal to frequency ν multiply by Planck's constant h ($h = 6.626069934 \times 10^{-34}$ J s).

The concept of photon already implies the duality of light as both particles and waves. Because in classical physics, light and electromagnetic phenomena are just waves, and quantum physics sates that the energy contained in these waves is quantized and has a minimum value related to Planck's constant. Each piece of energy also implicitly means one "particle"! Therefore, both light and electromagnetic waves should be regarded as particles, i.e. photons. Afterwards, Bohr's atomic model linked the emission of light quanta with the energy of electrons.

Louis de Broglie (1892–1987), was a French physicist and aristocrat, who originally majored in history, discovered that physics was his interest and turned to the study of quantum mechanics. In 1924, he wrote an amazing doctoral thesis [9], which made remarkable progress for quantum mechanics. De Broglie extended Einstein's research on the light dualism to electrons and other physical particles, and proposed the concept of wave assigns to any non-zero mass particle a "de Broglie wavelength" that is inversely proportional to the particle's momentum. This new concept of wave-particle duality in any substance made it difficult for his advisor Paul Langevin (1872–1946) to accept so he sent de Broglie's paper to Einstein for comments. Einstein immediately realized the value of this paper, and he thought that De Broglie "has lifted a corner of the veil." The support of Einstein established the position of wave-particle duality in physics and inspired another physicist Schrödinger (1887–1961). Schrödinger thought, since electron have duality, let's establish a wave equation for it. Two years later, Schrödinger's equation [10] came out, ushering in a new era of quantum mechanics.

In 1927, American physicist Clinton Davisson (1881–1958) of Bell Labs and his student Lester Germer (1896–1971) discovered the diffraction of electrons for the first time in an experiment. Almost at the same time, English physicist George Thomson (1892–1975) also observed the diffraction of electron beams in polycrystalline film experiments. For confirmed the concept of de Broglie wave, Davidson and Thomson were awarded the 1937 Nobel Prize in Physics. It is enviable that George Thomson's father Joseph Thomson (1856–1940) won the Nobel Prize in Physics in 1906 for his discovery of electrons. This contribution of "father and son" to electronic research is truly unprecedented.

In addition, the well-known double-slit electron interference experiment is also an excellent example of the electron-wave-particle duality. The electron must be regarded as a kind of wave and described by a wave function that satisfies the Schrodinger equation (or the Dirac equation under the conditions of relativity) in order to explain the double-slit experiment, because only waves can produce interference phenomena.

In Newtonian mechanics, the state of a particle at a certain moment can be described by its position and momentum in three-dimensional space. After the introduction of the wave function concept in quantum mechanics, the state of a single electron affects the entire space. Generally speaking, the state of a quantum system is called a "quantum state", usually is a "superposition state". The meaning of "superposition" is that: particles seem to be in two or more places or states at once. This is the peculiar quantum phenomenon that people usually use "Schrodinger's cat" to describe. That is, "Schrodinger's cat": the cat is simultaneously alive and dead.

It is the wave nature of electrons that leads to a series of puzzling quantum phenomena not found in classical physics.

2.2 Quantum Tunneling Effect

In quantum theory, even if the height of the barrier is greater than the energy of the particle, microscopic particles can penetrate or traverse the barrier with a certain probability, causing the "quantum tunneling effect". In classical mechanics, this is impossible. This phenomenon can be explained by wave functions of electrons in quantum theory, because according to wave dynamics theory, there is no impenetrable barrier.

The first application of tunneling effect was a mathematical explanation for alpha decay, which was done in 1928 by a Russian-American physicist George Gamow (1904–1968). This was one of the earliest achievements in quantum mechanics research on atomic nuclei.

In classical mechanics, there can be no weird thing like "through the wall", and it is impossible for a particle to cross a higher energy barrier than its energy. The potential barrier is like a mountain blocking the road. For example, we push a stone to reach a slope. If the slope is small, the kinetic energy of the stone is greater than the potential energy of the mountain, and it can be passed by. However, if the mountain is high, when the kinetic energy of the stone is less than the potential energy, the stone will stop halfway through, and it is impossible to pass. This is shown in the right part of Fig. 2.1.

For another example, we are listening to a lecture in a classroom with closed doors and windows. No one can go outside through the wall. But let us imagine that we and the classroom are becoming smaller and smaller… We have become alpha particles, and the classroom has become an atomic nucleus that prevents alpha particles from leaving. At this time, the situation is different. According to quantum theory, alpha particles in the tiny world have no fixed positions and are a fuzzy group of "wave packets." However, this wave packet is different from the "material wave packet" that we generally call diffuse, but a stable probability wave packet. Therefore, each of us was originally like clouds and fog permeating the entire classroom, even including the outside of the classroom, there are also our faint shadows. Just as the British physicist RH Feller listened to Gamow's lecture on "Tunneling Effect" at the Royal

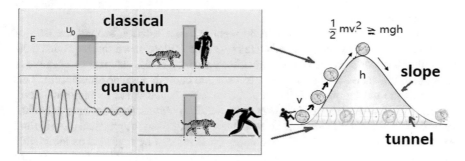

Fig. 2.1 Classical barrier and quantum tunnel

Society of London that winter, and said with a smile: "Anyone in this room has a certain chance to leave the room without opening the door!".

According to the wave theory, the wave function of electrons will permeate the entire space, and particles will appear at every point in the space with a certain probability, including points outside the barrier. In other words, the probability of a particle passing through the potential barrier can be solved from the Schrodinger equation. In other words, even if the energy of the particles is less than the energy of the barrier threshold, some particles may be bounced back by the barrier, but some particles will still pass through with a certain probability, as if there is a tunnel at the bottom of the barrier, as shown in Fig. 2.1.

The tunneling effect not only explains many physical phenomena, but also has many practical applications, including tunnel diodes commonly used in electronic technology, and scanning tunneling microscopes used in basic scientific research in laboratories.

We will see in later chapters that the carriers (including electrons and holes) in graphene follow a special quantum tunneling effect, so when they encounter impurities, they will not produce backscattering., Instead of jumping forward over the barrier with a theoretical 100% pass rate, this is the reason for the super local conductivity of graphene and the high carrier mobility.

2.3 Spins

For photons, spin may be the quantum–mechanical counterpart of the polarization of light; but for electrons, the spin has no classical counterpart. Therefore in general, we think spin as purely a unique concept only found in quantum theory. Although people often compare spin to rotation in classical physics (such as the earth), this analogy is only available to a certain extent. The rotation in the classical mechanics is the rotation of an object with respect to its center of mass, which is very different from the nature of spin. In other words, spin is an intrinsic property of microscopic particles, corresponding to a quantum number that can be an integer or a half integer. This intrinsic characteristic is fundamentally different from the classical rotational angular momentum.

Why can't the spin of a particle be regarded as a rotation around the axis of rotation? Because if you follow this analogy and calculate based on the value of the possible radius of the electron, the imaginary surface of the electron must move at more than the speed of light in order to generate enough angular momentum. This is against the theory of relativity. In addition, even if you try to maintain a certain "rigid ball" picture of electrons, it is hard to imagine photons. In quantum mechanics, photons and other elementary particles all have spins. So far, the physical theories and experiments have believed that elementary particles can be regarded as indivisible point particles. Point particles cannot rotate around their axes. Therefore we regard spin of particles as a kind of intrinsic angular momentum, a quantized property that cannot be changed.

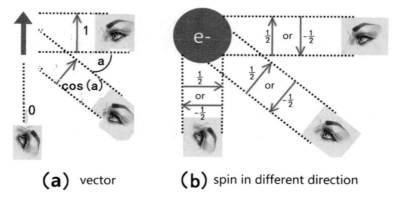

(a) vector   **(b)** spin in different direction

Fig. 2.2 Vector and spin (1/2)

Particles with half-integer spins are called fermions (such as electron spin is ½), and integers are called bosons (such as photon spin is 1). Composite particles also have spins, which are obtained through the addition of the spins of the constituent particles (which may be elementary particles); for example, the spins of protons can be obtained from the spins of quarks.

Fermions, such as electrons, have many quantum characteristics that completely do not follow the classical laws.

For example, the angular momentum in classical physics is a vector in three-dimensional space. We can observe this vector in different directions to get different projection values. As shown in the Fig. 2.2a, the upward red arrow. When we observe it from the right, its size is 1; when viewed from below, the projection value is 0; and when viewed from a certain angle α, Then get the value of $\cos(\alpha)$ continuously changing from 0 to 1 with angle α.

In quantum mechanics, the spin is quantized. No matter from which angle you observe the spin of electron, you can get only one of two values: ½ or $-$ 1/2. This behavior is quite different from vector case and as shown in the Fig. 2.2b.

The "up" and "down" states are the eigenstate of spin for electrons, which can be shown in the Fig. 2.3a. Most of time, the electron is in a superposition state where two states coexist.

Therefore, the electron spin is not a vector in the usual three-dimensional space, but can be regarded as a vector in a two-dimensional complex space. In other words, its operation law can be classified as "spinor". In a sense, the spinor can be regarded as the "square root of a three-dimensional space vector". However, this sentence still sounds difficult to understand. Where does the square root of the vector come from?

For example, a vector in a two-dimensional space can correspond to a complex number, so we may be able to understand the "square root of a vector" from the square root of the complex number. A complex number can be represented by its absolute value (modulus) and angle known as the argument of the complex number. If the square root of this complex number is required, it can be obtained by taking

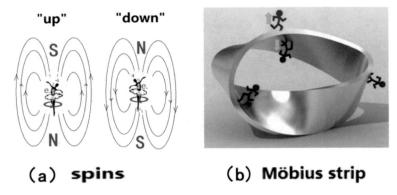

Fig. 2.3 The nature of electron spin

the square root of its modulus and halving the argument. Therefore, the argument of the square root of a complex number is half of the original complex number. When a complex number $(1, 0)$ rotates around the origin in the complex plane, that is, when it returns to its original value after $360°$, its square root only rotates half a circle ($180°$), staying at the position of the original vector in the opposite direction, only when the original complex number rotates two times around the origin, the square root complex number returns to the original position.

The spin of electrons has similar properties. When the spin makes a circle in space, instead of returning to its original state, it changes up to down, and down to up, just like the runner in Fig. 2.3b after moving around on the Mobius belt. It became the same as head down. It can also be seen from Fig. 2.3b that if the head-down runner continues its Mobius trip, after another round, it will turn head-up and return to its original up-state. It can be seen that this property of electron spin is exactly similar to the "vector square root" property described above.

2.4   Identical Particles

According to quantum theory, the particles are governed by wave functions that give the probability of finding a particle at each position. As time passes, the wave functions tend to spread out and overlap. It is impossible to accurately "track" particles, so it is impossible to distinguish them from each other. Therefore, quantum mechanics believes that the same kind of microscopic particles are "identical" and indistinguishable. Identical particles can be classified into bosons and fermions. These two types of particles follow different statistical laws: bosons obey Bose–Einstein statistics, and fermions obey Fermi–Dirac statistics. In the standard model of elementary particles, the protons, neutrons, and electrons that make up the structure of matter are all fermions, and the four particles, which mediate forces, including photons, gluons, etc., are all bosons.

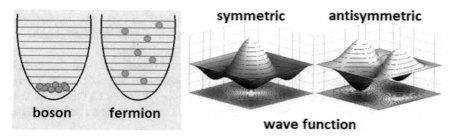

Fig. 2.4 Bosons and fermions

The different statistical properties of different microscopic particles are derived from their different spin wave functions and different symmetry caused by these functions. Bosons are particles with integer spins, such as photons with spin 1. The wave functions of the two bosons are exchange symmetric. In other words, when the roles of the two bosons are exchanged, the total wave function remains unchanged. Another type of particles called fermions has a half-integer spin. For example, the spin of an electron is one-half. The wave function of a system composed of two fermions is commutative antisymmetric. In other words, when the roles of two fermions are exchanged, the total wave function of the system only changes the sign, as shown in Fig. 2.4. Whether the wave function is symmetric or anti-symmetric, it will not affect the probability obtained after squaring, but it will affect the statistical properties of the two types of particles.

The two statistical laws apply not only to elementary particles, but also to composite particles. For example, protons, neutrons, and various types of mesons formed by the combination of quarks, and atomic nuclei formed by the combination of protons and neutrons are all composite particles., For composite particles, if they are composed of an odd number of fermions, they are fermions; if they are composed of an even number of fermions, they are bosons.

The concept of boson fermions defined according to statistical laws can also be extended to "quasi-particles" in solids and condensed matter.

In a semiconductor, electrons are affected by the nucleus and other electrons. The behavior of electrons can be regarded as free electrons with different masses, i.e. the effect of nucleus can be represented by mass-change. We call them "quasi-electrons", as well as "holes", they are not real particles either. These two types of quasiparticles (quasi-electrons and holes) are fermions. However, quasiparticles may also be bosons, such as Cooper pairs, phonons, etc.

Multiple bosons can occupy the same quantum state at the same time, and two fermions cannot occupy the same quantum state at the same time. This is a very important statistical difference between two. In other words, bosons are a group of friends; fermions are mutually exclusive independent. If a group of bosons go to live in a motel, they would like everyone to live in the same room, and one big room is enough; but if a group of fermions go to live in a motel, they need to provide each of them with a separate small room.

Electrons follow the "Pauli Incompatibility Principle" and are arranged in layers in atoms. This explains the periodic table of elements and describes the relationship between the chemical properties of matter and its atomic structure.

Because bosons like everyone living in the same room, everyone desperately squeezes to the lowest energy state. For example, a photon is a kind of boson. Therefore, many photons can be at the same energy level. We can get a super-intensity beam like a coherent state of laser that "all photons have the same frequency, phase, polarization, and direction of travel".

The different statistical behaviors of bosons and fermions as described above are also one of the most mysterious aspects of quantum mechanics!

2.5 Quantum Entanglement

The superposition of a single particle's wave function is an intriguing quantum phenomenon. When the concept of superposition state is applied to a quantum system of more than two particles, weird and strange phenomena will be produced. One of them is the "quantum entanglement".

The original concept of quantum entanglement was Einstein's hypothetical thought experiment, the EPR paradox [11], because he opposed the Copenhagen interpretation of quantum mechanics. Later, Schrödinger named it quantum entanglement. In 1964, the British physicist John Bell (1928–1990) proposed the Bell experiment and Bell's theorem, which made it possible to have a clear experimental method to verify the physical meaning of quantum entanglement. The original intention of EPR, Schrödinger, and Bell for quantum entanglement was to prove the possible inconsistency or incompleteness in quantum mechanics, and to use specific experiments to verify the localization hidden variables behind quantum theory.

The EPR article has been published for more than 80 years, regrettably, the results of many experiments did not support the "hidden variable" view of Einstein and others. On the contrary, the conclusions of the experiment proved the correctness of quantum mechanics time and time again. Despite the differences, the mechanism of quantum entanglement still needs to be explored; most physicists believe that this kind of counter-intuitive "ghost-like over-distance action" does exist.

What quantum entanglement describes is the high correlation between two electronic quantum states. This kind of correlation is not available to the classical particle pair, and is a unique phenomenon that only occurs in quantum realm, in which the underlying cause goes back to the wave nature of electrons. Readers may wish to imagine: two "probability wave packets" permeating space are entangled, obviously more "difficult to distinguish" than two "small balls" entangled together.

Consider a two-electron quantum system and use electron spin to explain "entanglement". Electron spin has only two simple eigenstates "up and down", similar to the positive and negative sides of a coin toss, unlike position or momentum, there are countless eigenstates.

For example, if the spin of two entangled particles is measured separately, and one of them gets the result as "up", then the spin of the other particle must be "down", if one of them gets the result as down, then the other spin of the particle must be up. Use a simple classic example to illustrate: Isn't that like dividing a pair of gloves into two boxes? One stays at A and the other gets to B. If you see that the glove at A is right-handed, you can know that the glove at B must be left-handed, and vice versa. No matter how far A and B are separated, even if they are separated into two planets, this law will not change.

To take it a step further, for a real glove, open box A to see that it is a right hand, close and open it again, it is still a right hand. Any time box A is opened, you will see the right hand and it will not change. But if the box contains electrons instead of gloves, you will not always see (observe) a fixed spin value, it may be "up" or "down". There is no definite value, up and down, both are possible) with a certain probability. Before the measurement, electrons are in a state of "up and down" superposition, a "life and death" state similar to the "Schrodinger Cat". Before the measurement, the state is uncertain, and after the measurement, "up" or "down" is known. Prior to the measurement, our "human" observers could not predict the result, the B electron, far away in the sky, somehow seem to be able to "perceive" the outcome of A electron being measured in advance, and adjusted to the state opposite to that of the A electron. In other words, no matter how far away the two electrons are, they seem to be able to achieve state synchronization. How is this possible? Moreover, if the synchronization of A and B's electrons is interpreted as the ability to exchange messages between them, the speed of this message transmission is very fast, and it has greatly exceeded the speed of light. Does this violate the theory of relativity?

The phenomena and experimental facts described in the previous sections are basically recognized by all quantum experts. There are a wide variety of different interpretations: in addition to the more mainstream Copenhagen interpretation, there are also multi-world interpretations, ensemble interpretations, transaction interpretations, and so on. But there does not seem to be a theory that can explain all experiments and satisfy all physicists. This is where Einstein was not satisfied with quantum mechanics. In the following sections and subsequent chapters of this book, we use the so-called mainstream view of quantum mechanics: the Copenhagen interpretation.

2.6 Wave Functions

The role of Schrödinger equation in quantum mechanics is similar to Newton's second law in classical mechanics. However, the difference is that Newtonian once brought clarity the physics world, but quantum mechanics after Schrödinger's equation added more confusion.

The solution of Newton's equation is the trajectory of particles in space with time. This trajectory is tangible and easy to understand. Even the invisible trajectory can be imagined most of the time. But the solution solved from Schrödinger's equation is a "wave function" that permeates the entire space! This wave function is very useful, explaining that experiments have developed theories, but what exactly is it? How to relate it to the motion of small spherical electrons in people's minds?

Schrodinger first thought: Does the wave function represent the density of charge? This idea did not work intuitively, and failed miserably in calculations. In 1926, Born (1882–1970) gave a probability explanation, assuming that the square of this wave function represents the probability of electrons appearing at a certain point in space. This idea seemed to have successfully explained the physical significance of wave functions. However, Schrodinger himself does not agree with this statistical explanation. After that, a series of quantum bizarre phenomena and interpretations began with the wave function, including Heisenberg's uncertainty principle, Bohr's complementary theory, collapse of wave functions, the meaning of quantum measurement, and quantum Entanglement, etc., which made the famous Bohr–Einstein debates about quantum mechanics for many years. The physicists are basically divided into two major groups: the Copenhagen group represented by Bohr, and the opposition led by Einstein, Schrodinger and others. This led to the so-called "quantum century war" between Einstein and Bohr. The effects have continued to these days.

Of course, Einstein did not oppose quantum mechanics itself, nor did he oppose the probability theory, he just could not accept Copenhagen's probabilistic interpretation of wave functions. The best Einstein did was use counterexamples to propose a few thought experiments, he himself failed to create a constructive and new quantum theory framework and interpretation.

On the contrary, the Copenhagen Institute of Theoretical Physics led by Bohr at that time became the world's quantum research center, including Born, Heisenberg (1901–1976), Pauli (1900–1958) and Dirac (1902–1984). A group of young people with the same age as quantum mechanics are the main members of Niels Bohr Institute, and they have made outstanding contributions to the establishment and development of quantum mechanics. The Copenhagen interpretation has long dominated the physical world and is a widely accepted mainstream view. Even if it may be replaced by other interpretations or theories in the future, the Copenhagen institute and interpretations are indispensable in the development of quantum mechanics.

2.7 Uncertainty Principle

The uncertainty principle says that the position and momentum of a particle cannot be determined at the same time. The more precisely the position of a particle is given, the less precisely can one say what its momentum is, and vice versa.

In fact, before Schrödinger derived the Schrödinger equation, Heisenberg and Bohr had established the first mathematical foundation for quantum mechanics: matrix mechanics. Afterwards, Schrödinger proved that the two descriptions of matrix mechanics and the wave dynamics of Schrödinger's equation are mathematically equivalent. However, physicists are accustomed to differential equations, because they are used in Newtonian mechanics. People also like intuitive wave function graphs and dislike the boring mathematical operations of matrix mechanics. Even though the physical meaning of the wave function is not very clear, scholars enthusiastically studied and applied Schrodinger's equation, while leaving matrix mechanics aside. This has made Heisenberg quite disappointed. He is determined to give his own theory a more intuitive picture.

Heisenberg tried to use images to describe the trajectory of electrons, but found that electrons actually have no trajectories, because the position and momentum of the electron cannot be determined at the same time. For example, the best way to measure the electronic position is to use a laser whose wavelength is smaller than the range of movement of the electron. The electrons in the atom have a range of motion of only 10^{-10} m, but the possible speed of motion is very large, 10^6 m per second. In this fast-moving situation, the electrons are hit by the laser photon, the speed and position are constantly changing, and no accurate value can be measured.

Heisenberg believes that there is no fundamental reality that the quantum state describes, just a prescription for calculating experimental results. Heisenberg's uncertainty principle was inspired by Einstein's thought of "observable" quantities, according to Einstein; a perfect theory must be based on directly observable quantities. But, Einstein would not accept this until his death. Heisenberg uncertainty principle was:

$$\Delta x \Delta p_x \geq \hbar$$

In the above inequality, perhaps the lower limit of the right side of the inequality sign is not completely correct, but it does not affect the basics of this principle. The uncertainty principle is a mathematical theory in nature. It determines the bound to be restricted by the so-called regular conjugate variables appearing in pairs in mathematical equations. The mutually restricted conjugate quantities (pairs) are not limited to position and momentum. Energy and time, time and frequency in signal transmission, etc. are all examples of conjugate variable pairs.

2.8 Wave Function Collapse

While proposing the uncertainty principle, Heisenberg also proposed another central concept of the Copenhagen interpretation: wave function collapse, the purpose of which is to explain the relationship between the uncertainty principle and quantum measurement.

Physics is only concerned with observable things. However, observation requires measurement, and the measurement of electronic behavior inevitably allows electrons to interact with some external influence. In this way, the measurement of observing electrons must be accompanied by interference of equipment, as shown in Fig. 2.5.

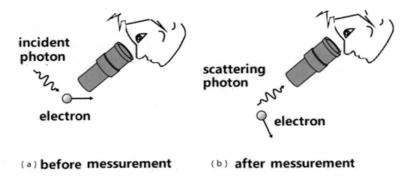

(a) **before messurement**   (b) **after messurement**

Fig. 2.5 Measurement affects the motion of electron

For classical measurement behavior, the scale of interference is much smaller than the measurement scale and can be ignored, but it cannot be ignored in quantum measurement. Therefore, the microscopic world has to follow the uncertainty principle.

The position of a microscopic particle with a certain momentum is uncertain, and it is not known where it is. Once we look at it, it instantly appears in a certain position, so a certain value of the position can be obtained. In order to solve this contradiction, Heisenberg introduced wave function collapse. Heisenberg said that because collapse of the "wave function", electron's originally uncertain position collapses into a certain position due to human observation. This concept was later popularized by mathematician John von Neumann and incorporated into the mathematical formula of quantum mechanics.

To describe the wave function, we introduce the concept of quantum superposition state. The motion of electrons can be expressed as the superposition of different definite position states, or as the superposition of different definite velocity states. The wave function is the superposition coefficient. When measuring the position, the quantum state randomly "collapses" to a quantum state with a fixed position; when measuring the velocity, the quantum state randomly "collapses" to a state with a fixed speed, i.e. measurement led superposition state collapses to a certain state. The probability is related to the superposition coefficient, which is the square of the magnitude of the wave function.

In other words, quantum mechanics uses two processes to describe electrons. One is the evolution of the wave function described by the Schrodinger equation (or Dirac equation), which is reversible; the other is irreversible caused by measurement, which is called "wave function collapse". The former is recognized by most people, and the latter belongs to the Copenhagen interpretation. Even today, the issues raised by the wave function concept have not yet been satisfactorily answered. It is said that Bohr himself did not fully accept the idea of wave function collapse.

2.9 Probability

In both classical physics and quantum theory, the term "probability" is used to represent the uncertainty of events, but their physical interpretations are quite different. Probability can be defined as a description of the uncertainty of things. According to the concept of classical physics, it is believed that probability occurs because people have insufficient knowledge. From the point of view of quantum mechanics, uncertainty does not come from lack of knowledge, but belongs to the inner nature of matters.

In the framework of classical physics, uncertainty comes from our lack of knowledge, because we do not have enough information, or there is no need to know so much. For example, when a person throws a coin upwards and catches it with his hand, the direction of the coin seems to be random, and it may be upward or downward. But from the perspective of classical mechanics, this randomness is because the movement of the coin is not easy to control, so that we do not understand (or do not want to know) the detailed information of the coin flying out of the hand. If we know exactly the force on each point when the coin flies out, and then solve the mechanical equation, we can completely predict the direction it will fall. In other words, classical physics believes that behind the uncertainty, there are some undiscovered "hidden variables". Once they are found, any randomness can be avoided. We can say, hidden variables are the source of probability in classical physics. This is exactly what Einstein meant when he said, "God doesn't throw dice!" Einstein didn't understand quantum probability, but stubbornly believed that God's dice were rolled according to the laws of deep "hidden variables", and thus proposed the famous EPR paradox.

The uncertainty in the quantum theory explained by the Copenhagen group is different. They believe that the uncertainty in the microcosm is inherent and essential. There is nothing to hide deeper variables, and some are just "the wave function collapses" to a fixed value with a certain probability.

The electron double-slit experiment confirmed that an electron passes through two slits "at the same time" and its more bizarre behavior is manifested in the measurement!

In order to explore how interference occurs in the electron double-slit experiment, the physicist put two particle detectors in the two slits of the double slit experiment, trying to measure which slit each electron has gone through? How is interference fringes formed? Strangely, once you want to observe which slit the electron passed through by any method, the interference fringes disappeared immediately, and the wave-particle duality seemed to vanish. The experiment gave the same result as the classic bullet experiment!

How to explain this kind of quantum paradox theoretically? The Copenhagen group believes that electrons in the microscopic world are usually in an uncertain superposition state that cannot be described by classical physics: it is this and that. For example, when the electron before the measurement reaches the slit, it is in a certain (positional) superposition state: both at the slit position A and at the slit position B. After that, "Each electron passes through two slits at the same time!", an interference phenomenon occurred.

Once the electrons are measured in the middle, a "wave function collapse" occurs in the quantum system. The wave function that originally represents the uncertainty of the superposition state collapses to a fixed eigenstate. In other words: the collapse of the wave function changes the quantum system, making it no longer the original quantum system. Once the quantum superposition state is measured, it returns to the classical world according to certain probability rules.

This interpretation brings many problems (other interpretations have other problems). The Copenhagen interpretation directly confuses people: how to understand the nature of measurement? Who can measure? Can only "people" measure? Where is the boundary between measured and unmeasured?

American physicist John Wheeler (1911–2008) argued that: "no elementary quantum phenomenon is a phenomenon until it is a registered ('observed,' 'indelibly recorded') phenomenon." This tongue-twister-style passage leads people to question Copenhagen in this way: Does the moon only exist when we look back? This question is actually a misunderstanding of Copenhagen's interpretation.

Classical physics has always believed that the research object of physics is the objective world independent of the existence of "means of observation." Copenhagen's interpretation of quantum mechanical measurement seems to blend the subjective factors of the observer into the objective world. The view that subjective and objective are inseparable in measurement does not mean denying the existence of the objective world. In addition, there is still a lot of dilemma about quantum phenomena. For more information about quantum mechanics, please refer to Ref. [12].

Chapter 3
Microscopy Graphene

From the introduction in Chap. 1, we already have a basic understanding of graphene. As illustrated in many publicity pictures, graphene is nothing but a large net woven into a regular hexagon (honeycomb shape) of carbon atoms. In academic terms, it is called "quasi-2D crystal structure". The words and images of regular hexagons, crystals, and 2D (Two-dimensional) nets all refer to the positional relationship between atoms. Where are the electrons in graphene? In addition, although graphene is just a net, it is a super-large, super-thin and super-strong net. Each grid is a perfect hexagon, and each knot is a carbon atom. Since this net is only one atom thick, the height is negligible, only length and width. When 3 million such "mesh pieces" are superimposed on top of each other, the result is only 1 mm thick. Therefore we refer to graphene as a 2D structure. In the real physical world, "Two-dimensional" material does not exist. For graphene, with depth of a single atom and tightly spaced atoms on the x–y dimension what happens on the graphene network can be explained by the laws of quantum mechanics introduced in Chap. 2.

In Chap. 4, more theoretical discussion on the physical properties of general crystals will be covered. In this chapter, the author will first lead the reader to the microscopic world of graphene and take a closer look at how the carbon atoms of graphene "hold hands" with each other. How is the electronic cloud distributed? Then, from the perspective of experimental detection, see what methods scientists use to detect and observe the atoms and electrons in a two-dimensional lattice such as graphene, and take a closer look at graphene from a microscopic point of view.

3.1 Atom and Electron Cloud

In general the atoms of any matter are composed of nuclei and electrons. The interpretation of atomic models is quite different in classical physics and modern physics.

© Guangxi Science & Technology Publishing House 2022
T. Zhang, *Graphene*, https://doi.org/10.1007/978-981-16-4589-1_3

In classical mechanics, the trajectory of electrons is a "curve line" in space, that is, the solution of Newton's equation of motion. In quantum mechanics, because of the uncertainty principle and wave-particle duality, the concept of orbital is meaningless. The state of electrons follows the wave equation in quantum mechanics. The solution of this type of equation is a "wave function" that permeates the entire space. Sometimes, we still use the intuitive image of "orbits", but we must remember in our minds that quantum theory makes two basic corrections to classical orbits: first, the "orbit" is quantized; second, the orbit is not a single curve but is actually an electron cloud permeating the space around the atom, the density of which represents the probability of electrons appearing.

Considering the "stationary state" that has nothing to do with time, quantum and classical images are still very different. The classical stationary state of an electron is a single point in space with a certain position and momentum, while in quantum mechanics, to describe a stationary state of electron, the wave function of the entire space is needed. The solution of Schrodinger equation is usually called "quantum state". Different quantum states correspond to different energy, orbital angular momentum, and spins. Solving the time-independent Schrödinger equation is a more basic quantum mechanics problem. If an electron in a quantum state has a certain energy E, we usually say that the electron is at the energy level of E.

In fact, the steady-state wave function is also a function of time. The square of the wave function, that is, the probability density, is independent of time. The wave function itself is not the observable, but the probability density is the observable. In other words, for steady state, the probability density and all observables do not change over time.

An important achievement after the establishment of the Schrödinger equation in quantum mechanics is the analytical solution of the wave function of the hydrogen atom. The energy level of the hydrogen atom and the electrons of the hydrogen atom under different combinations of quantum numbers are accurately calculated. The shapes of electron cloud of hydrogen atom have been displayed in Fig. 3.1. The method can be extended to other more complex multi-electron atomic systems. All the effects on the electrons in the system are approximated on average, and the Schrodinger equation under this average potential field is solved. This established an accurate (or approximate) mathematical model for complicated atoms and explained the periodic table well.

Although the wave function is a function that permeates the entire space, different wave functions still have different shapes. In Fig. 3.1, the probability density of electrons is displayed, which is the square of the wave function. Near the nucleus, the electron density is high, and far away, the probability of electrons appearing is small and the density is thin. Therefore, the electron orbits of atoms in quantum mechanics are described by images with different densities, the electron orbits became a group of clouds of different shapes that accumulate around the atoms. That is why people call them "electronic clouds."

The atomic number of carbon is 6, and its nucleus consists of 6 protons and 6 neutrons. There are 6 electrons around the nucleus, including 2 electrons in the first

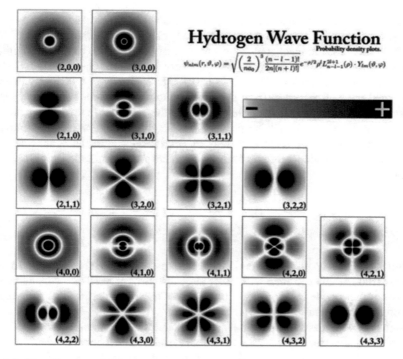

Fig. 3.1 Wave function of hydrogen atom (from Wikipedia)

layer and 4 outer valence electrons. According to the classic atomic planet model, the electron rotates around the nucleus, as shown in Fig. 3.2a.

According to the quantum mechanics, electrons have no clear orbits, but only appear in a certain space area to a certain probability to form an electron cloud.

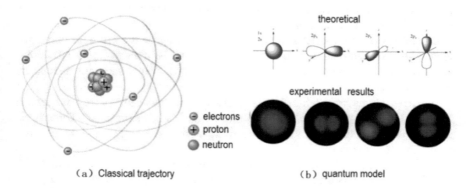

(a) Classical trajectory   (b) quantum model

Fig. 3.2 Carbon atom structure

The s orbital corresponds to a centrally symmetric electron cloud, and the p orbital corresponds to an axisymmetric electron cloud. As shown in Fig. 3.2b [13].

Using the classical mechanics, one can roughly estimate the order of magnitude of the electron velocity in the atom, which is very large. If the speed of light is set to 1, then the speed of electrons will be about 0.5–0.8. Such a large speed is limited to a very small range. Because of the uncertainty principle, we cannot accurately detect electrons on the orbit. Is it possible to observe the electronic cloud using modern technology in the laboratory? The answer is yes, it can also be simulated by computer. The various microscopy techniques that will be introduced in the later sections of this chapter have made it possible for humans to observe atomic and even subatomic structures.

3.2   Covalent Bonds and Hybrid Orbitals

Figure 3.2 depicts an electron cloud in the structure of a single carbon atom. In most cases, carbon atoms combine with other atoms to form molecules, or many carbon atoms combine with each other by themselves, for example, to form the various carbon allotropes that we described in Chap. 1.

Matter is composed of molecules, and molecules are composed of atoms. How possible atoms are form stable molecules (or crystals)? The theory based on quantum mechanics. By sharing electron pairs between atoms in a molecule, each atom has a stable electronic structure. The study of chemistry at the molecular level is the field that is most affected by quantum mechanics in addition to physics itself. It is precisely because of the application of quantum mechanics to the chemistry of valence bond theory and molecular orbital theory that the formation of covalent bonds in molecules is explained. Chemists have figured out the nature of atoms combining to become molecules, that is, the nature of chemical bonds. The interaction between chemical bonds is also within the category of four basic forces summarized by physics. The dominant role in the molecular composition is electromagnetic force, without any additional so-called "chemical force."

Chemical bonds are electrical in nature. According to the different ways and degrees of this electrical action, chemical bonds can be divided into ionic bonds, covalent bonds and metal bonds. The atoms in most compounds are bonded by covalent bonds.

When two atoms approach each other, there will be attract and repel forces between particles. How can they "hold hands" to form a stable covalent bond? Three basic principles must be observed: "electron pairing", "lowest energy", and the electron cloud "overlaps the most."

For example, Fig. 3.3 illustrates the above-mentioned "three major principles" in terms of spins. Electron pairing means that when two atoms form a bond, each provides an unpaired electron, and the two electrons spin in contrary directions. If two electrons have opposite spins, the potential energy curve is shown in Fig. 3.3a, so that the energy of the diatomic system can be reduced to form a bond. When electrons

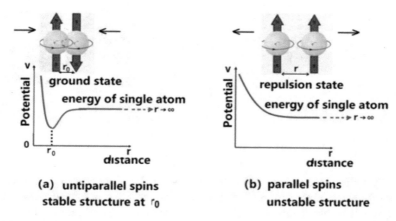

Fig. 3.3 The effect of spins on molecular bonding

are close to each other, the energy decreases, which is lower than the energy of a single atom, causing mutual attraction until the distance between nuclei is equal to r_0. It is the state with the smallest potential energy of the system, called the "ground state" of the diatom. Figure 3.3b describes the situation where the spin directions of two electrons are the same. At this instance, the closer the two electrons are, the greater the energy and the inability to form a valence bond, which is called the "repulsion state". When forming a bond, the more the two atomic orbitals overlap, the denser the electron cloud between the two nuclei, and the stronger the covalent bond formed. This is called the principle of maximum atomic orbital overlap.

One question remains: How can we achieve maximum overlap? There are two basic ways for electron clouds to overlap. People compare them to "head-to-head" and "side-by-side", corresponding to two types of covalent bonds: σ bond and π bond, as shown in Fig. 3.4.

Each carbon atom has four electrons capable of bonding. Therefore, carbon atoms can form chemical bonds in many ways.

The four electrons in the outer layer of the carbon atom include one 2s orbital and three 2p orbitals. But in a crystal composed of carbon atoms, these four orbitals

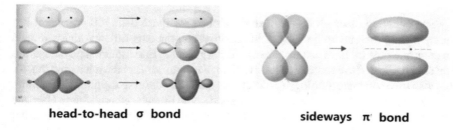

head-to-head σ bond   **sideways π bond**

Fig. 3.4 Two types of electronic cloud overlap when forming a bond

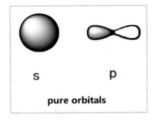
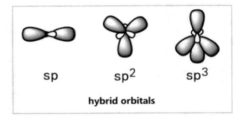

Fig. 3.5 sp hybrid orbital

generally redistribute energy and determine the spatial direction to form four new hybrid orbitals, as shown in Fig. 3.5. With the help of different hybridization modes (sp, sp^2, sp^3) between carbon atoms, crystal structures with different physical and chemical properties can be formed.

Hybrid orbitals were proposed by American chemist Linus Pauling (1901–1994) in 1931. It is still the basis of modern valence bond theory in essence; it has enriched and developed in terms of bond forming ability and molecular spatial configuration. In bonding, several atomic orbitals with similar energies in the same atom are linearly combined to redistribute the energy and determine the spatial direction to form an equal number of new atomic orbitals. This process is called hybridization and the new orbitals are called hybrid orbitals.

The chemical bond formed by the hybrid orbital is strong and the bond energy is large, which makes the generated molecule more stable. The shape of angle distribution diagram changes when the orbit is hybridized. Generally, one end is large and the other is small. The large end can form a larger overlap, making the hybrid orbit more directional. For example, in diamond, the redeployment of the 2s and three 2p orbitals of carbon atoms produce four new identical, sp3 hybrid atomic orbitals, forming a firm regular tetrahedral structure.

3.3 A Snapshot of the Inside of an Atom

The internal structure of the atom is now well known, it can be observed in the laboratory.

The quantum theory that has been developed for more than 100 years still has many basic problems to be solved. In recent years, the experimental fields related to quantum physics have achieved extraordinary results. Experimental physicists are not only dedicated to overcoming the problem of how to connect the micro and macro, but also use modern technology to directly or indirectly observe the atomic structure. The 2012 Nobel Prize in Physics was awarded to the French scientist Serge Haroche and the American scientist David J. Wineland for "ground-breaking experimental methods that enable measuring and manipulation of individual quantum systems". The two scientists realized the work of imprisoning individual quantum systems in the laboratory and have provided new investigational tools for researchers and established a foundation to understand the relationships between matter and light at the smallest of scales.

In 2013, Dutch researchers announced that they captured the world's first atomic structure. In fact, the problem is still how to observe the quantum phenomena occurring on the micro scale on the macro scale. They claimed that they had taken pictures of the atomic structure by observing the wave function of electrons in the atom.

What Dutch scholars call the "first atomic structure diagram" is the diagram of the hydrogen atom. The hydrogen atom is the simplest atom. After the Schrodinger equation was established, it was first used to describe the hydrogen atom and it was a great success. When using quantum mechanics to deal with atomic energy levels and solving wave functions, hydrogen atom is the only situation in which quantum mechanics can get accurate analytical solutions.

The method used by researchers at the Max-Born Institute in the Netherlands (A. S. Stodolna, etc.) was proposed by Russian theoretical physicists (V. D. Kondra-tovich and V. N. Ostrovsky) about 30 years ago. Their experimental setup is shown in the Fig. 3.6 [14].

Kondratovich et al. proposed an experimental method, which is equivalent to establishing a microscope system, and predicted that it can be used to observe the wave function of electrons. In the experiment they designed, the hydrogen atoms ionized by the laser are placed in an electrostatic field in the z-direction, and the hydrogen atoms will be in an excited Stark quantum state.

The wave function of the Stark state can be written as the product of two wave functions. Because it contains a fixed electrostatic field, the Schrödinger equation of the hydrogen atom can still be solved by the method of separating variables. But at this time, the original spherical symmetry of the hydrogen atom is destroyed. Therefore, a linear coordinate transformation is required to convert r (the distance between the electron and the nucleus) and z (the displacement of the electron along the electric field axis) into the so-called "parabolic coordinates". In parabolic coordinates, the

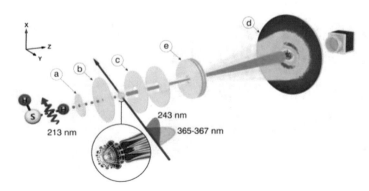

Fig. 3.6 Experimental setup of Dutch researchers (https://physics.aps.org/featured-article-pdf/10.1103/PhysRevLett.110.213001)

Schrodinger equation can separate variables and solve the wave function of the Stark quantum state.

This idea from thirty years ago is very fascinating, but it is not easy to experiment. Since there are no stable hydrogen atoms in chemistry, lasers are used to dissociate other molecules first, and then the electrons of hydrogen atoms are excited to high-energy states with lasers. As shown in the figure above, the photoelectrons emitted by the atom, carrying the information of the Stark quantum state wave function, are magnified by three electronic lens elements and projected to a two-dimensional detector placed about half a meter away and perpendicular to the electrostatic field. An interference pattern is generated and this pattern corresponds to the wave function of Stark's quantum state. These wave functions are standing waves, and the nodal pattern of the wave function reflects the quantum number of the state. Because of the coordinate transformation, the original three quantum numbers (n, l, m) of the hydrogen atom are also transformed into three other quantum numbers (n1, n2, m). Here, $n = n1 + n2 + m + 1$.

The experimental results are shown in Fig. 3.7.

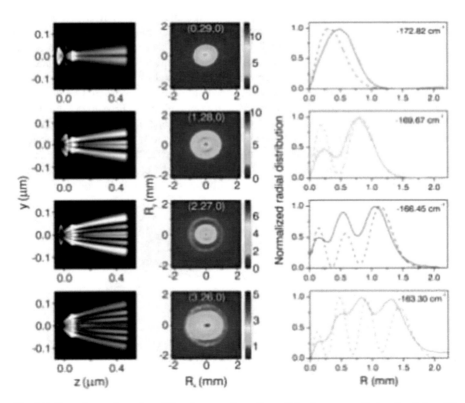

Fig. 3.7 Experimental results of Dutch researchers (https://physics.aps.org/featured-article-pdf/10.1103/PhysRevLett.110.213001)

The figure above corresponds to four Stark quantum states. The quantum numbers (n1, n2, m) are respectively (0, 29, 0), (1, 28, 0), (2, 27, 0) and (3, 26, 0). Electrons in hydrogen atom are excited to the high-energy state: $n1 = 0, 1, 2$ and 3. Here n1 is not large, but n2, 26–29, is relatively large. The wave function shown in the middle color picture has 1, 2, 3, and 4 nodes, which can be clearly seen from the picture. The maximum number of curves in the figure on the right is also consistent with the number of nodes.

Through this experiment, the wave function image of the Stark state of the hydrogen atom electron is enlarged to a millimeter-level size, which can be observed with the naked eye. The magnification is about 20,000 times.

3.4 Microscopy Techniques

Technology has grown by leaps and bounds over past decade. The graphene research has benefited a lot from this. Because of the excellent special properties of graphene structure, we need to use some special experimental methods to detect it.

3.4.1 Optical Microscope

Although our naked eyes cannot see atoms, a microscope can, which is extension of human vision. Therefore, in order to examine the structure and morphology of graphene samples, microscopy is undoubtedly the preferred method.

Microscopes have a long history. The first optical microscopes were created by Hans Jansen and his son Zacharias Jansen in the Netherlands in 1590. But the optical microscope has its diffraction limit. Diffraction limit means that the resolution of the microscope measurement is limited by the wavelength, which is not difficult to understand intuitively. In fact, the so-called measurement process is the process of comparing the measurement tool with a certain corresponding physical quantity of the measured object. To give a simple analogy, when we use a ruler to measure the size of an object, the scale on the ruler should be much smaller than the size of the object, or at least comparable to accurately read the value of a certain length being measured. The relative size of the ruler and the scale of the measured object will affect the accuracy of the measurement. The same is true for the process of detecting objects with light waves. Light waves are the "ruler" used for microscopic measurement, and its wavelength is like the scale on the ruler. Light of a certain wavelength, or to put it more broadly, electromagnetic waves of a certain wavelength, is only suitable for detecting objects of the corresponding size, as shown in Fig. 3.8.

When wavelength of the detection wave is much larger than the scale of the object, the wave can only go around, which means that the "diffraction limit" of the microscope using this wave is reached, as shown in Fig. 3.9.

EM wavelength range

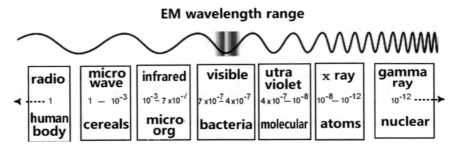

Fig. 3.8   Electromagnetic spectrum and the range of the measured object

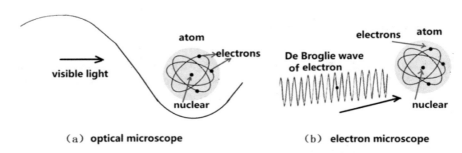

Fig. 3.9   The resolution of the microscope is limited by the wavelength

An optical microscope that uses visible light expands the functions of the human eye, which is equivalent to magnifying the object more than a thousand times. The diffraction limit limits its magnification to around 1600 times. The wavelength range of visible light is approximately in the range of 390–700 nm, which is the same order of size of bacteria, so it is suitable for biological research. In other words, bacteria are most suitable for observation with a visible light microscope. Because radio waves with large wavelengths have already "detoured" when passing through bacteria; and for bacteria there is no need to use other types of microscopes with very short wavelengths, "Why use a butcher's knife to kill a chicken"?

It can be seen from Fig. 3.8 that the scale of the atomic nucleus is about 10^{-12} m (0.001 nm), which is 4–5 orders of magnitude different from the wavelength range of visible light. Therefore, it is impossible to observe the internal structure of the atom with an optical microscope. The invention of the electron microscope solved this problem.

According to the principle of quantum mechanics, the de Broglie wavelength of an electron is related to the electron momentum p. The larger p is, the smaller the wavelength, which can be calculated according to the following formula: wavelength = h/p, where h is Plank constant.

Generally speaking, the de Broglie wavelength of electrons is much smaller than the wavelength of light waves. For example, the wavelength of 100 kV electrons is about 0.0037 nm (nanometers), maybe it can be used to detect atomic structure and observe electron clouds, see Fig. 3.8.

3.4.2 Electron Microscope

In the 1920s, quantum theory was established, and in 1924 de Broglie proposed the matter wave conjecture. At that time, physicists had the idea of designing and manufacturing electron microscopes. Biologists hope to overcome the diffraction limit of optical microscopy and use electron microscopy to help them distinguish pathogens such as viruses. If the electron beam is really a wave similar to light, there will be interference and diffraction phenomena. The structure of an optical microscope can be imitated to make an electron Microscope. Electron beam emitted from electron gun can be applied to the sample, then be focused using EM field, and finally form image.

To successfully develop an electron microscope, step-by-step experiments are needed to support it. For example, we must first use experiments to prove that electrons are indeed waves!

In 1926, a German physicist Hans Busch (1884–1973) suggested the first magnetic electron lens. In 1927, Davidson and Germer, and Thomson, discovered the diffraction of electron. This confirmed the concept of de Broglie waves and strengthened people's confidence in the development of electron microscopes.

In 1931, German physicists Ernst Ruska (1906–1988) and Max Knoll developed the first transmission electron microscope (TEM) based on Busch's idea of EM lens (Fig. 3.10a). In 1938, Ruska continued to research at Siemens and made the first commercial electron microscope. Forty-eight years later, Ruska won the Nobel Prize in Physics in 1986.

In addition to the transmission electron microscope, there is also a scanning electron microscope (SEM), see Fig. 3.10b.

In the transmission electron microscope, the electron stream accelerated by the high voltage penetrates the sample and directly obtains the projection of the sample on the screen. The design idea of TEM is simple, but the disadvantage is that the sample must be cut into very thin slices. For thicker samples, a higher voltage is required to accelerate electrons.

The structure of the scanning electron microscope is different. The electron beam does not have to penetrate the sample; rather it is focused on a small part of the sample and then scans the sample line by line. The electron beam used for scanning is incident on the sample, causing secondary electron scattering on the sample surface, and the microscope observes these secondary electrons to obtain information on the sample surface. Because the electrons in the SEM do not have to transmit through the sample, the voltage for accelerating the electrons does not have to be very high.

It is also possible to take a combination of the two: let the electron beam pass through the sample and scan it. This type of microscope is called a scanning transmission electron microscope (STEM). The first STEM was launched in 1937.

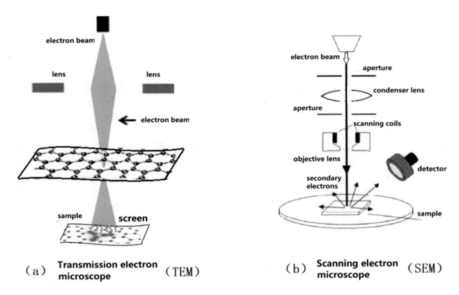

(a) **Transmission electron microscope** (TEM)   (b) **Scanning electron microscope** (SEM)

Fig. 3.10 Schematic diagrams of electron microscope

The electron microscope successfully overcomes "diffraction limit" of optical microscopy and enabled proof of quantum mechanics: electrons are fluctuating.

The electron microscope at that time still had many shortcomings. Once instance, in revealing the surface structure of an object, it will have problem if the energy of the electron beam is too large or too small. If the speed of the electron is too high, it will penetrate into the depths of the material and there is no secondary emission. Otherwise, if the speed of the electron is too low, it is easily deflected by the electromagnetic field of the sample, making it difficult to obtain ideal observation results.

3.4.3 Scanning Probe Technology

Scanning probe technology includes scanning tunneling microscope and atomic force microscope. How can the surface properties of objects be accurately detected without causing any damage to the sample surface? In 1981, scientists Gerd Binnig and Heinrich Rohrer at the IBM Lab in Zurich took a different approach. Instead of using an electron gun to emit electron beams, they cleverly use the tunneling phenomenon between the scanning probe and the sample surface and rely on receiving tunneling current to detect the surface of the object.

Their method is called Scanning Tunneling Microscope (STM) because it uses the quantum tunneling effect to probe the surface structure of substances. The principle is shown in Fig. 3.11a: Keep a layer of insulating medium between the metal probe

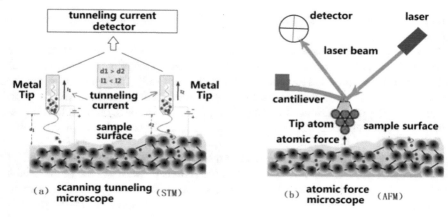

Fig. 3.11 Surface detection microscope

and conductive samples, the insulating layer is thin enough to allow electrons to pass through and generate leakage current, or tunnel current. Then, let the probe move along a certain reference plane in all directions of the sample surface, and scan it like a SEM. The difference is that the scanning tool of the SEM is a focused electron beam, and the STM uses a metal probe. If the surface of the sample is uneven, the thickness of the insulating layer between the two conductors will change, which will change the tunnel current detected. In other words, the change in the magnitude of the tunneling current received reflects the tiny fluctuations of the surface being detected, and the tunneling current carries information about the surface morphology of the sample. Therefore, it is possible to observe the single-atom level fluctuations on the surface and send the current signal to the computer for processing, and then an image of the sample surface can be obtained. Scanning tunneling microscope allows us to see individual atoms on the order of nanometers.

Scanning tunneling microscope is widely used in surface science, material science, life science and other fields because it can directly observe the atomic structure of the surface of the object. It has opened up many new research fields and has become the key nanotechnology. The two inventors of the scanning tunneling microscope shared the Nobel Prize in Physics in 1986 with Ruska, the inventor of the electron microscope.

STM and later invented atomic force microscope (AFM) are both surface structure analysis instruments. The advantage is that the resolution is very high, up to one angstrom. Compare to SEM, resolution of STM is one order of magnitude higher. STM can also analyze insulators, biological samples, solid–liquid interfaces, etc. The price of equipment is relatively low. The disadvantage is that it can only analyze the surface structure and topography. The magnification cannot be easily changed like an electron microscope. The scanning time is also very long. Although it has a special purpose, it cannot replace SEM, TEM, STEM. Among current microscopes, the highest resolution is still TEM. It can be less than one angstrom, down to 0.8 angstrom.

The atomic force microscope (Fig. 3.11b) uses the high sensitivity of its cantilever to work. The cantilever can really "feel" the mechanical and electrical properties of the material, so when it is applied to graphene, not only the topography can be measured, but also the structure, mechanical and electrical properties. AFM is good at providing high resolution below 10 pm. Graphene is a single layer of carbon atoms, and many electromechanical characterization techniques can rely on the cantilever of AFM.

The application of electron microscope overcomes the diffraction limit of optical microscope, but the limit represented by wavelength is not the only factor that limits the resolution of the instrument. In fact, the resolution of the manufactured instrument depends on many other factors, such as the influence of spherical aberration and chromatic aberration of magnetic lens used in the electron microscope, etc. For scanning tunneling microscopes and atomic force microscopes, the resolution also depends on the size of the probe and the method of controlling the position of the probe when it is moving.

3.5   Spectral Analysis

Spectroscopy is a method for analyzing the structure of the sample based on the interaction between the graphene sample and the incident wave. After the incident wave is scattered or absorbed by the atomic structure in the sample, its wavelength and intensity will change. Through the analysis of the spectral changes, we can get information about the sample.

Spectroscopy includes Raman spectroscopy, infrared spectroscopy, ultraviolet–visible spectroscopy, X-ray, photoelectron spectroscopy, nuclear magnetic resonance, etc., as an example, here is a brief introduction to Raman spectroscopy.

The principle of Raman spectroscopy is Raman scattering, which was discovered in 1928 by the Indian physicist Sir Chandrasekhara Raman (1888–1970), for which he won the Nobel Prize in Physics in 1930. The Chandrasekhara family of India has two Nobel Prize winners in physics, the other is Raman's nephew Subrahmanyan Chandrasekah (1910–1995), the latter won the Nobel Prize in Physics in 1983 for his early discovery of the Chandrasekah limit related to stellar evolution and black hole formation.

There are two ways for photons to collide with matter particles: elastic scattering and inelastic scattering. The frequency of light waves does not change after elastic scattering, such as Rayleigh scattering; the frequency of light waves changes after non-elastic scattering, which is called the Raman Effect, or Raman scattering. When a photon collides with a material particle (or quasi-particle), if the photon loses energy, the frequency of the scattered light will be lower than that of the incident light, which is called Stokes scattering; another type of inelastic collision is During the collision, the material particles release energy and the photon energy increases, so the frequency of the scattered light is higher than the frequency of the incident light, which is called anti-Stokes scattering, see Fig. 3.12

Raman scattering is very weak. Generally, the spectrum formed by combining Rayleigh scattering and Raman scattering is called Raman spectrum.

For graphene device research, it is important to determine the number of graphene layers and the effect of defects. Experimental results have proved that Raman spectroscopy is a simple and reliable method to characterize the above two characteristics of graphene.

Sensitive to the structure of matter, its high spectral resolution, high spatial resolution, and non-destructive analysis make Raman spectroscopy a standard and ideal analysis tool in the graphene field. The Raman spectrum of graphene is characterized by several peaks, such as the G peak and 2D peak as shown in Fig. 3.13. For different

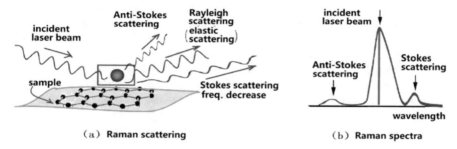

(a) **Raman scattering**   (b) **Raman spectra**

Fig. 3.12 Schematic diagram of Raman scattering and spectral shift

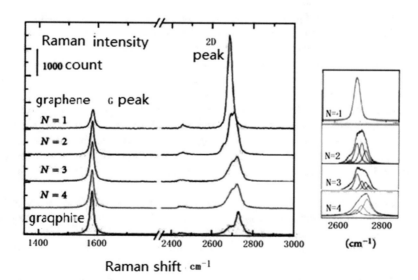

Fig. 3.13 Raman spectra of graphene with different layers

layers of graphene, the Raman frequency shifts of these two peaks are different. From this, the number of layers of the sample can be determined.

3.6 Angle-Resolved Photoemission Spectroscopy

The spectroscopy method introduced in the previous section observes the spectrum of incident light waves scattered from a solid, and the changes in the position of the spectral lines to different experimental samples. A light wave of a certain frequency is emitted from the light source to the sample, and then the scattered wave carrying the solid structure information is received.

If the energy of the incident light wave is high enough, above a certain threshold, the work function of the material, the energy of the photon can be absorbed by an electron. Then, the electrons near the surface can leave the sample within one billionth of a second, escape the metal surface and become free electrons, forming a photocurrent. This phenomenon is called the photoelectric effect.

The photoelectric effect is widely used. Here is a method of using it to detect atomic structure: angular resolved photoelectron spectroscopy (ARPES). This method was once hailed as: a microscope that can see the electronic structure.

For graphene and condensed matter physics, the magic of angle-resolved photoelectron spectroscopy is that it can be used to display and study the electronic energy level structure of solids, to observe the energy band diagram of solids. The energy band diagram is different from the atomic structure diagram inside matter. The observation of the atomic structure map shows the real atoms in real space. For example, the graphene network structure taken by the electron microscope is an enlarged image of the real space position of the carbon atoms in the graphene. The energy band diagram does not exist in real space. It describes the relationship between electron energy E and wave vector k, which is drawn along various directions in the Brillouin zone of wave vector space. In other words, the abscissa of the energy band diagram is the wave vector (or momentum), not the position in space.

For example, Fig. 3.14a, b are the theoretical prediction and ARPES observation results of a certain topological insulator energy band diagram, respectively. The M and Γ in the figure are vectors in the wave-vector space. After we introduce "Lattice and Band" in Chap. 4 and Topological Insulator in Chap. 6, readers will have a deeper understanding of Fig. 3.14. This section only briefly describes the principle of ARPES.

As mentioned earlier, the principle of ARPES is the photoelectric effect. In the usual photoelectric effect, light is used to eject photoelectrons, and then the photocurrent is received somewhere. The ARPES method is to accept and carefully analyze the energy–momentum relationship of these photoelectrons in different directions, so it is called "angle-resolved photoelectron spectroscopy."

ARPES uses light with very high energy, such as synchrotron radiation, and then measures the energy and momentum of the photoelectrons. It measures the speed of electrons (corresponding to the electron kinetic energy E_{kin}) and the angle of

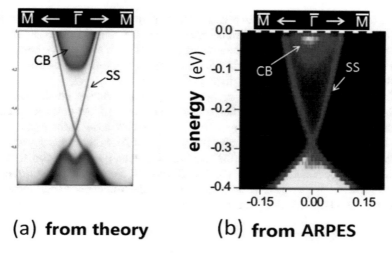

Fig. 3.14 Energy band diagram and ARPES output diagram

emission (θ, ϕ). As shown in Fig. 3.15, the collecting lens and electron energy analyzer calculate the number of free electrons with a certain kinetic energy within a small solid angle range. The corresponding set of measured values (E_{kin}, θ, ϕ) are transferred to the computer, and the calculation program, based on the measured data, takes into account the law of conservation of energy and the law of conservation of momentum at the same time, and the binding of electrons in the sample before

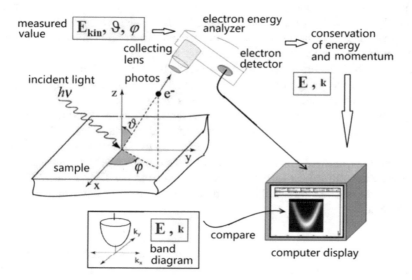

Fig. 3.15 Working principle of ARPES

being excited can be calculated in reverse. This functional relationship (E, k) reflects the dispersion properties of electrons inside the crystal sample material, which is the so-called band diagram. At the same time, ARPES can also get the energy state density curve and momentum density curve, and directly give the solid Fermi surface. Finally, the calculation results are displayed in images.

ARPES has been widely used in various important fields of condensed matter physics. It has played a very important role in the research of low-dimensional material structure, high-temperature superconductivity, and material surface state. These applications in turn have also promoted the progress of ARPES technology and other auxiliary equipment, such as the combination of a technology like Mott scattering that can distinguish electron spins into ARPES, so that it can obtain the third important information of the material in addition to the electron energy and momentum: electronic Spin has played a great role in the study of topological insulators.

3.7   Experimental Observation of Graphene

The various experimental detection methods described above are essential for studying the structure and performance of graphene. The following images are some examples of results.

Example 1:

Graphene itself has certain wrinkles and is not absolutely flat. If there are defects in the graphene lattice, it will affect its electronic, chemical, magnetic, or mechanical properties. Graphene has many defects in its structure. For example, the structure of ideal graphene is equiangular hexagon, but there may be some pentagons and heptagons, which constitute the defects of graphene. In addition, there may be vacancies in the crystal lattice, impurity doping, edge effects, grain boundary rotation, layer accumulation or distortion, etc., which are collectively called defects.

The various techniques introduced earlier in this chapter can help researchers determine the shape and type of defects, so as to determine what defects a certain preparation method will cause. How does this defect affect the function of graphene? How should the manufacturing process be improved to reduce defects?

The right picture of Fig. 3.16 is a photo taken by the Oxford University graphene research team in 2013 with a transmission electron microscope, and the left is a schematic diagram of the atomic structure drawn from the TEM results. There is a carbon atom space in the middle of the picture, making the hexagon above the space a pentagon.

Example 2:

The ideal graphene is a single layer, and the preparation process does not necessarily obtain a single layer material. Therefore, a single layer assembly with less than 10 layers is generally called graphene. What is the relationship between graphene properties and number of layers? How many layers are there? These are problems

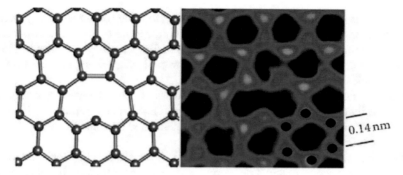

Fig. 3.16 Defects in the graphene lattice [15]

that researchers need to solve. There are many ways to determine the number of layers. Raman spectroscopy is a simple and easy method to measure the number of layers. In addition, the layer number can also be determined using atomic force microscopes and transmission electron microscopes (Fig. 3.17).

An optical microscope can quickly and easily measure the number of layers, and it is not destructive to the sample, but it is only suitable for graphene prepared on a substrate with obvious contrast differences. Raman spectroscopy is fast and effective, with high resolution, and does not damage the sample. It is suitable for measuring AB stacking (lattice shift) graphene, but it is difficult to distinguish AA stacking (lattice isotopic) graphene. Atomic force microscope is a direct and effective method, but the observation range is small, the efficiency is low, and the accuracy is affected by many factors. The number of layers seen by TEM is simple and intuitive, but it may damage the sample.

Example 3:

With the help of various detection technologies, it can be said that atoms have been "see". IBM also claimed to be able to manipulate and move individual atoms to make an "atomic movie". In recent years, there have also been reports of experiments in

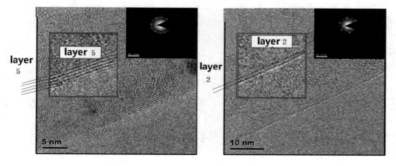

Fig. 3.17 Determine the number of layers of graphene

which "electronic clouds" have been observed. However, scientists always expect
to continue to penetrate into the interior of a single atom to see more information
about the "orbits" of electrons. Graphene is a thin film with only one layer of atomic
thickness and may be a suitable candidate for exploring the subatomic level and
observing electron clouds.

Transmission electron microscopy is a good way to observe its atomic structure,
but not easily used like a mobile phone camera. TEM usually requires complicated
calculations and processing, or need to be supplemented by other physical processes
to achieve a special purpose.

The energy-filtered transmission electron microscope (EFTEM) shown in
Fig. 3.18b was used to obtain a clear image of each electron cloud in a single atom
of graphene, Fig. 3.18a.

The energy-filtered transmission electron microscope uses the filter lens group to
select energy after the electron beam passes through the TEM. The magnetic field
only allows electrons with energy in a narrow window to pass through, and filters
out the excess electrons, leaving only Electronics that carry useful information. The
above results were obtained in 2016 by an international research team led by Professor
Peter Schattschneider of the University of Vienna.

Example 4:

Figure 3.19 shows the band structure of graphene obtained using angle-resolved
photoelectron spectroscopy (ARPES). On the left is the theoretically predicted band
structure of graphene. Two cones are connected upside down. The cones are called
Dirac cones, and the point where they meet is called Dirac point. The energy band
diagram is used in solid-state physics to describe the relationship between the various
energy levels that an electron may have and its momentum. We will introduce the
basic concepts of crystals and energy bands step by step in the next chapter.

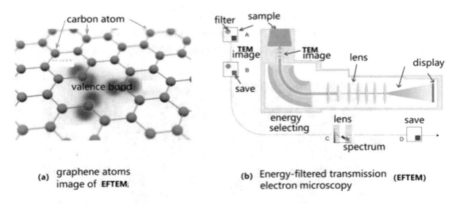

(a) graphene atoms image of EFTEM.

(b) Energy-filtered transmission electron microscopy (EFTEM)

Fig. 3.18 Covalent bonds of graphene observed [16]. Credit: Image courtesy of Vienna University
of Technology

Fig. 3.19 Band structure of graphene

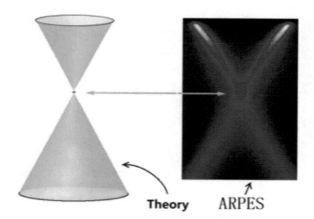

Theory   ARPES

Chapter 4
Lattice and Energy Band

4.1 Structure Determines Properties

The special property of graphene come from its unique structure and thus determines the application where graphene can be used. The macroscopic properties of a material should be explained by its microstructure and the relationship between its structure and the surrounding environment. This is why we cannot directly see the electrons inside the atom, but we can build various theoretical models for its motion. The purpose of these models is to explain the various characteristics of the material in the macroscopic measurement. Quantum mechanics is also such a microscopic theory. It was established to explain black body radiation and photoelectric effect. Now, it helps us understand graphene correctly.

The hybrid orbitals were introduced in the previous chapter. The outermost layer of a carbon atom has 4 electrons, which can form 4 covalent bonds with the electrons of other atoms. Due to the different hybridization methods in the process of forming covalent bonds, different materials are formed. Diamond and graphite are two typical examples. The four electrons around the diamond carbon atom have uniform contact forces in all directions, forming sp^3 hybrid orbitals, which are strong σ-bonds, so they form a strong force in three-dimensional space. The connected tetrahedral skeleton shape makes the diamond have high hardness. Among the 4 electrons around the carbon atoms of the graphene material, only 3 form sp^2 hybrid orbitals, that is, strong σ–bonds. They combine the carbon atoms into a sideways-shape to form the 2-dimensional lattice structure of graphene. The one electron left from each carbon atom in graphene wanders around the upper and lower 2-dimensional lattice, which are the common electrons of the plane lattice atoms. The electrons form a weak π–bond. When graphene layers are stacked to form a graphite material, it is easy to slide between layers, and the strength of the σ–bond on the plane is masked by the sliding of this weak π–bond. When the graphite is separated into a single layer of graphene, the strength of the three σ–bonds in the 2-dimensional lattice is exposed.

© Guangxi Science & Technology Publishing House 2022
T. Zhang, *Graphene*, https://doi.org/10.1007/978-981-16-4589-1_4

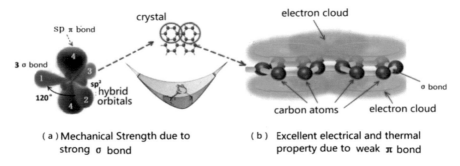

Fig. 4.1   The mechanical and electrical properties of graphene and hybrid orbitals

As shown in the left picture of Fig. 4.1a in graphene, there are three sp^2 hybrid orbitals (120° each other) forming from one s and two p electrons. These three σ–bonds in the hexagonal lattice make graphene with the highest elastic modulus and strength.

Experiments have shown that the spring constant of graphene was in the range 1–5 N/m, and its strength is 100 times higher than the best steel in the world. At the same time, it has excellent flexibility and elasticity. It is one of the materials with the best mechanical properties found so far. It can be bended and the stretching range can reach 20% of its own size. In other words, it is currently the thinnest and strongest material in nature. As for the strength of graphene, if you use a piece of graphene with an area of 1 m^2 to make a very thin and light hammock, the bed itself weighs only 1 mg, but it can bear a one kilogram Cat, see Fig. 4.1a on the right.

Due to the special structure of graphene, in addition to mechanical properties, it also exhibits properties that general materials do not have in other aspects. For example, single-layer graphene has ultra-high light transmittance. The reason for this is obvious because it is originally composed of only one layer of atoms. The experimental results show that the absorbance of single-layer graphene in a wide wavelength range is only 2.3%, that is, the light transmittance reaches 97.7%. For the same reason, the graphene material has a large surface area, the ratio of the area of the material to its mass, up to 2630 m^2/g.

Three of the four outer electrons of each carbon atom contribute to the lattice and determines the mechanical and optical properties of graphene. The free electron, or the one that's leftover, becomes a public electron in graphene that affects the materials electrical and thermal conductivity properties. The law of movement of this electron is determined by the crystal and band structure of graphene, which is the main content of this chapter.

4.2 What Is a Crystal?

Atoms, ions and molecules are arranged in space according to a certain periodicity to form a solid with a specific regular geometric shape, which is called a crystal. Therefore, the most essential feature of crystals is periodicity. Natural or artificially synthesized solid materials can be divided into three categories: crystals, quasi-crystals and amorphous materials. Most of the solid materials in nature are crystals, such as alum, snowflakes, etc. Glass and rubber are example of amorphous materials. Quasi-crystals are between crystalline and amorphous, and are solid materials manufactured in the laboratory.

The concepts related to the periodic structure and properties of crystals include "Bravais lattice, Bragg reflection, Bloch wave, Brillouin zone", etc., respectively named after several physicists related to crystal research.

4.2.1 Bravais Lattices

The Bravais lattice was discovered by French physicist Auguste Bravais (1811–1863). Bravais studied possible lattice structures from a geometric point of view, and established a lattice model of crystals. The types of crystal lattice geometry are limited. There are only five types of Bravais lattice on a two-dimensional plane. Oblique lattice, square lattice, hexagonal lattice, rectangular lattice, and centered rectangular lattice, as shown in Fig. 4.2.

The arrangement of three-dimensional crystals is more complicated and can be summarized into seven major crystal systems. Each crystal system is related to fourteen spatial lattices; refer to the schematic diagram in Fig. 4.3 for additional details.

Graphene is a two-dimensional crystal, which one of the five two-dimensional Bravais lattices in Fig. 4.2 represents it? At first glance, it looks more like a diamond lattice, but it is not exactly the same. The hexagon of the diamond lattice has one more atom than that of graphene. Scientists have proved that the lattice structure of graphene belongs to a kind of compound lattice, consisting of two sets of Bravais diamond lattices, and there is a translation between the two sets of diamond lattices, as shown in Fig. 4.4a in A (White) and B (black).

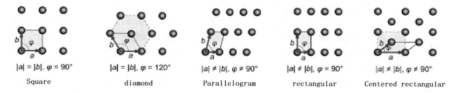

$|a| = |b|, \varphi = 90°$	$|a| = |b|, \varphi = 120°$	$|a| \neq |b|, \varphi = 90°$	$|a| \neq |b|, \varphi = 90°$	$|a| \neq |b|, \varphi \neq 90°$
Square	diamond	Parallelogram	rectangular	Centered rectangular

Fig. 4.2 Two-dimensional Bravais lattices

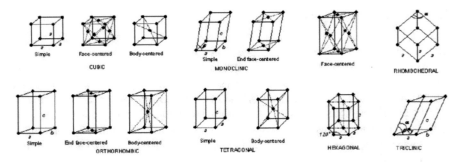

Fig. 4.3 Three-dimensional (fourteen types) Bravais lattice

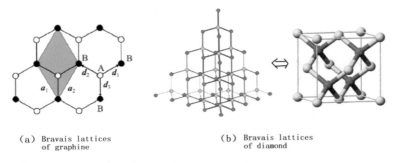

(a) Bravais lattices
      of graphine

(b) Bravais lattices
      of diamond

Fig. 4.4 Example of complex lattice crystal structure

Diamond is an example of a three-dimensional compound lattice, which is formed by two sets of face-centered cubic Bravais lattices translated along the diagonal of the body and then stacked. The distance of translation is a quarter of the diagonal of the body, as shown in Fig. 4.4b.

4.2.2 Bragg Reflection

British physicist Sir William Henry Bragg (1862–1942) and his son Sir William Lawrence Bragg (1890–1971) used experimental methods to detect crystal space and establish diffraction theory for the lattice. In the end, the father and son pair shared the 1915 Nobel Prize in Physics. This was the only time a father and son took the Nobel podium to receive the prize. Moreover, little Bragg was only 25 years old at the time, making him the youngest Nobel winner to date.

Bragg and his son used electromagnetic waves to detect crystal lattice and pioneered crystal structure analysis. They used x-rays, electron waves, neutron waves, etc. to study various crystal structures and established a theoretical basis. Figure 4.5 is a schematic diagram of the crystal reflection law discovered by Bragg. It can be

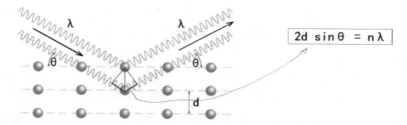

Fig. 4.5 Bragg reflection in crystal

seen from the figure that for a certain incident angle θ, if the optical path difference between two waves reflected from two parallel crystal planes at a distance of d is equal to an integer multiple of the wavelength λ, it meets the conditions for two waves to interfere with each other and strengthen: $2d\sin\theta = n\lambda$. This formula is called the Bragg equation. When the angle of the equation is satisfied, the light wave is strengthened. There is also the other scenario where the two waves interfere with each other and cancel each other. In this way, we can observe the diffraction image on the receiving screen and then analyze it further. Study the grid information carried by the image.

4.2.3 Bloch Wave

Bloch wave refers to the wave function of electrons in the crystal lattice.

In 1928, when Einstein, Bohr, and others were arguing about how to interpret quantum mechanics, a young man, one of the students of Werner Heisenberg, took a different approach to study the solid crystal lattice.

He is the Swiss-American physicist and the 1952 Nobel Prize winner Felix Bloch (Felix Bloch, 1905–1983). Bloch's contribution to solid-state physics is to solve the Schrödinger equation for electrons in the lattice, and to establish the electron band theory based on it.

The motion of electrons in the lattice is a multi-body problem, which is very complicated, but after Bloch made some approximations and simplifications, the conclusions are intuitive and concise. He studied the simplest one-dimensional lattice situation, and then generalized it to three-dimensional.

Bloch first solved the wave function and energy eigenvalues of free electrons (potential field is 0) in vacuum. Then, he treated the periodic potential field of the lattice that affects electrons as a perturbation, and thus obtained the approximate solution of the Schrodinger equation for the electrons in the lattice.

According to Bloch's conclusion, the wave function of electrons in the crystal lattice is similar to the wave function of free electrons in vacuum, but the amplitude of which is modulated by the periodic potential of the crystal lattice (see Fig. 4.6).

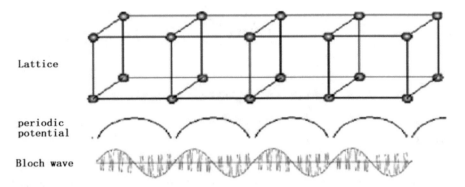

Lattice

periodic
potential

Bloch wave

Fig. 4.6 Schematic diagram of Bloch wave in the lattice

4.2.4 Brillouin Zones

The crystal structure is a periodic arrangement of atoms, and free electrons move in this periodic potential field. For periodic functions, physicists have a powerful mathematical tool, which is the Fourier transform. What are the advantages of Fourier transform? Let's take the Fourier transform of sound waves as an example.

The left picture of Fig. 4.7a shows the time signal of a certain frequency sound (for example, a monotone "do"). It is a series of intensity values that change periodically according to time. The time range starts from 0 and continues to the end of the sound. If you consider the Fourier transform of this signal, that is, the function curve in the frequency domain, as shown in the right figure of Fig. 4.7a, it is only a function limited to a small range. Only when the frequency is equal to around 500 Hz, the sound intensity is not zero.

The above example shows that the Fourier transform only extracts the frequency information of the sound and uses a simpler way to describe the sound signal that extends over the entire time range in the frequency space. The same is true for crystals.

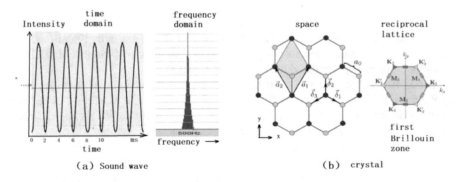

Fig. 4.7 Schematic diagram of Fourier transform and Brillouin zone

The "potential field" generated by the periodically arranged atoms in the lattice is a function that continues in the entire space. The concept of Fourier transform can be applied here to maximize the outcome. The spatial translation symmetry of the crystal removes redundant information and expresses the periodic potential field function in a concise way.

Physicists call the crystal points in real space "lattice" and the space after Fourier transform "reciprocal lattice", or wave vector (k) space. The relationship between the inverted grid and the grid is similar to the relationship between the frequency domain and the time domain. The potential function that continues in the whole space is confined to a limited range in the reciprocal lattice space. This range is called the "first Brillouin zone". The first Brillouin zone of graphene is a regular hexagon, as shown in the schematic diagram on the right of Fig. 4.7b.

4.3 What Is Energy Band?

As mentioned above, the k-space after Fourier transform can be used more effectively to study crystal problems. From the point of view of quantum mechanics, all particles or electromagnetic waves have wave-particle duality such that wave vector k is related to momentum p. Therefore, there is only a constant factor between k and p. When this book talks about wave vector space, k-space, reciprocal lattice, Brillouin zone, etc., all of them are similar concepts: it refers to the crystal structure after Fourier transformation. When doing specific calculations, you need to figure out exactly which space it is, but if you only want to understand the physical meaning, the series of nouns above do not need to be strictly distinguished.

The energies of the bands in solid state physics are calculated in "k-space". Energy band is an extension of the concept of energy level. According to quantum mechanics, the energy of electrons in an atom cannot be taken continuously, and can only take several discrete values, called energy levels. In a crystal, a large number of atoms are stacked together, some electrons are shared by all atoms, and the energy level expands into an "energy band." The energy band theory qualitatively clarifies the general characteristics of common electron behavior in crystals. Different materials have different energy band diagrams. Their characteristics illustrate the electron transport properties of materials, characterize the difference between conductors, insulators, and semiconductors, and explain the mean free path of electrons in crystals.

In the energy band diagram, the wave vector (k) space curve is generally used to characterize the electrons. We still start from the arrangement of atoms in real space, and first discuss how electrons are shared and how the energy band is formed. For simplicity, the following discussion will take a 1-dimensional lattice as an example.

From a microscopic observation, an atom is like a family, with the nucleus in the center surrounded by electrons. Multiple nuclei in the crystal are arranged into a lattice, and the electrons around them are divided into two categories: some are in a bound state and only move around the nucleus to form an electron cloud around the nucleus; some are free and shared by all nuclei. Free electrons wander around,

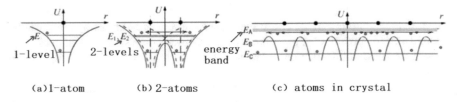

(a) 1-atom   (b) 2-atoms   (c) atoms in crystal

Fig. 4.8 From 1 atom to crystal, energy levels become energy band

forming a cloud of shared electrons. The number and distribution of free electrons in a solid substance determine the different conductivity of the substance. The energy band diagram of a certain crystal material describes the possible energy value of an electron with a certain wave vector.

Figure 4.8 describes the process of sharing electrons in the crystal. The top line represents the arrangement position of the atoms in the lattice, and below the line are the potential energy curves and electron energy levels of the atoms in the lattice.

The electrons in a single atom only move in the Coulomb potential field of its own nucleus, and its energy is determined by several separate energy levels. An energy level diagram similar to Fig. 4.8a can be drawn. The Coulomb potential in Fig. 4.8a looks like a well, which confines the movement of electrons in it. Electrons are a kind of fermions and don't like to live in groups. They live in layers around the atoms, stand in different levels, do not invade each other, and are in order. Starting from the lowest energy state, they line up to enter, occupying separate energy levels of each atom, that is, several horizontal lines in the single atom Coulomb potential well described in Fig. 4.8a. When two atoms are very close together, their potential wells merge as shown in Fig. 4.8b. From the outside, the double potential well still has a high Coulomb barrier, and electrons cannot go outside the double atom. But it's different inside the two atoms. The electrons with lower energy still live in their own "wells" properly; the electrons with higher energy (such as the energy E in Fig. 4.8a) become the shared electrons of the two atoms, but they are still required to have their own houses. Therefore, the original energy level E splits into two very close energy levels, E_1 and E_2.

In crystals, many of these identical atoms are connected together, as shown in Fig. 4.8c. At this time, the adjacent potential wells are all connected, and the number of freely running electrons increases. They are shared by all the nuclei of the entire crystal lattice and become free electrons in the crystal. Although the shared electrons are free, they still maintain the nature of "unwilling to live in groups". Each electron must live on a floor, that is, occupy a quantum state alone, thus splitting the energy level. If it is assumed that each atom contributes a free Electrons (as in the case of Fig. 4.8c), and if the total number of atoms in the solid is N, the original energy level is split into N energy levels. When N is very large, the energy levels of the splits are very close and seem to be continuous, forming an energy band.

A continuous energy band, such as the shaded part represented by E_A in Fig. 4.8c, includes many different energy levels (quantum states). These energy levels can be

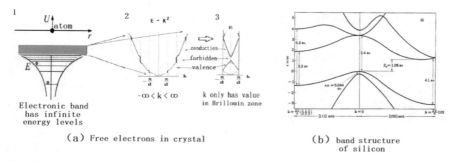

(a) Free electrons in crystal

(b) band structure
of silicon

Fig. 4.9 Energy band diagram

occupied by shared electrons in the crystal. An electron with certain energy value has a certain momentum value. The shadow in Fig. 4.8c represents the energy band in the real lattice space. Another way to express it is to put the energy levels in the shadow and expand according to their corresponding momentum, so that a curve in the momentum space will be obtained. For example, the energy of a free electron in a vacuum is proportional to the square of the wave vector k, that is, $E = k^2$. After being expanded in momentum space, it is a parabola extending from $k = -\infty$ to $+\infty$, see Fig. 4.9a.

Figure 4.9a shows the movement of electrons in a crystal. The wave function of the electron is modulated by the periodic potential field of the ion in the crystal lattice, so that the shape of the parabola (Fig. 2) of the electron energy band is locally destroyed. This loss mainly occurs on the boundary of the Brillouin zone, those k values that satisfy the Bragg diffraction equation. Far away from these boundary values, electrons can still be regarded as free electrons, conforming to the square (parabola) law, while near the boundary k value, electrons bound by the strengthened periodic potential field. The plane wave shape that carries energy spreading everywhere breaks the continuity of the original curve at the boundary of the Brillouin zone, and also separates the degenerate energy levels, resulting in a band gap, as shown in Fig. 4.9a (small Fig. 2).

In addition, if the translational symmetry is considered, the parabola can be folded repeatedly, and finally reduced to a limited energy band diagram of the Brillouin zone. Finally a typical energy band curves as shown in Fig. 4.9a (Fig. 3) is obtained: the range of k value is only from $-\pi/d$ to π/d. The energy band includes conduction band, forbidden band, and valence band.

The band structure of a real solid is much more complicated. For example, Fig. 4.9b shows the band structure of silicon.

So, what is a band structure? The points on the curve in diagram represent the possible states of the electron. Different materials have different band structures. The structure and shape of the energy band diagram, the position and width of the valence band, forbidden band, and conduction band determine the difference between insulators, conductors, and semiconductors.

4.4   Conductors, Insulators, Semiconductors

There are valence band, forbidden band and conduction band in the band structure. The forbidden band is an energy range that electrons cannot occupy. Below the forbidden band is the valence band, and above the forbidden band is the conduction band. The valence band has been filled with valence electrons, and the conduction band is generally empty. The electrons on the valence band cannot move freely because they are crowded with electrons. When there is enough extra force (light or heating, etc.) the electron can make the sudden jump over the forbidden band onto the conduction band. Once inside the empty conduction band, electrons run freely and become free electrons in the crystal.

As shown in Fig. 4.10, in an insulator, there is a wide energy gap (forbidden band) between the conduction band and the valence band. It is difficult for valence electrons to break through this gap to reach the conduction band. Therefore insulator cannot conduct electricity.

The conductor has no band gap, that is, $E_g = 0$, the conduction band and the valence band are connected together, or sometimes overlap each other. The electrons in the valence band can reach the conduction band and become free electrons shared by the entire solid. Therefore, the conductor has strong conductivity.

For semiconductors, they have conduction band, valence band, and gap. Similar to insulators, the valence band is full, but the gap Eg is very small. In this way, it is easy to undergo transition and conduction under external effects (such as light, heating, doping, etc.). Conductivity of those materials is generally much worse than that of conductors, so it is called a semiconductor.

There is also a horizontal dashed line labeled "Fermi level" in Fig. 4.10, which determines an energy value. We can use the analogy of electron occupying a "room" (energy level) to explain this.

Curves in band structure indicate rooms that the electron can live in. But they do not tell how electrons distributed in these rooms under certain conditions. There is a

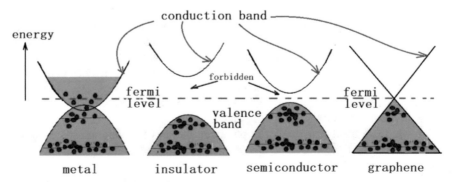

Fig. 4.10 Different band structures

parameter that tells us this information. This parameter is the Fermi level shown by the horizontal dashed line in Fig. 4.10.

When the temperature is close to zero, rooms are full below the Fermi level and empty above it. If the temperature rises a little, the situation is slightly different. The temperature rises and the kinetic energy of electrons increases. They are not as quiet as before and start to stir in the room, especially those electrons close to the Fermi level. The higher temperature, more likely the electron will jump successfully.

When the absolute temperature T is not 0, the Fermi level determines the probability distribution of the electron "in the room". The Fermi level position in Fig. 4.10 compares the difference between graphene and metal.

In summary, when the temperature rises, only electrons near the Fermi level can easily jump around to generate electric charge. This is the mechanism by which solids are conductive or non-conductive and determine various physical properties. Therefore, we are only interested in band structure near the Fermi level, because they determine the transport properties of electrons (or holes).

It can also be seen from Fig. 4.10 that the energy band structure of graphene is very special, different from the three previously mentioned. It looks a bit like a semiconductor energy band diagram, but there is no gap between the valence band and conduction band. If you compare the energy bands of graphene and metals, the difference is near the "Fermi level": the electron density of graphene at the Fermi level is 0, while that of metal is not 0. The detailed band structure of graphene will be discussed in the next chapter.

4.5 Free Electrons in Crystals

The energy band structure gives a clear and intuitive image of "steady state" for shared electrons in the solid. It explains what intrinsic energy values the electrons can have in the crystal periodic potential field. The next question is: If an external electromagnetic field is added to a solid, how will these electrons move? How is their law of motion different from that of free electrons in vacuum?

It is a very complicated problem to use quantum mechanics to study crystal electrons in the external field. Common electrons in the crystal not only feel the external force, but also the periodic field of the crystal at the same time. Under normal circumstances, the external field is much weaker than the crystal period potential. Therefore, semi-classical methods are generally used to study the movement of electrons in an external field, that is, the eigenstates of the electrons in the periodic field are first considered. They are the solutions of quantum mechanical wave equation under conditions of the crystal periodic potential.

The so-called "semi-classical" method means that the first step is quantum, and the second step is classical. First, use quantum theory to find the steady state solution of the electron in the periodic field. This steady state wave function can be regarded as an electron cloud surrounding the electron that reflects the probability of the electron's appearance. Then the electron moving in a crystal lattice with external

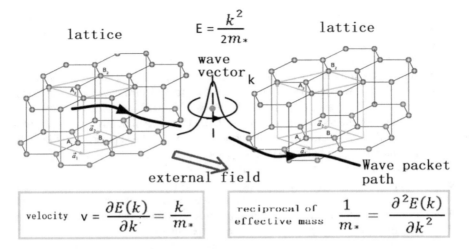

Fig. 4.11 Bloch electrons in the crystal

field can be similar to the movement of free electrons in vacuum with external fields. Both can be regarded as the classical movement of a "wave packet" in an external field.

What is the difference between the movement of free electrons in a crystal and in a vacuum? The difference lies in the mass of electrons. The mass of electrons in vacuum is equal to the inherent rest mass m_0, while the mass of electrons in crystals is the effective mass m^*. For the sake of distinction, we generally refer to free electrons in the crystal as Bloch electrons, as shown in Fig. 4.11. The effective mass m^* of the wave packet is generally not equal to the rest mass m_0, and may be larger or smaller. It is determined by the stationary wave function of the electron in the periodic potential field. That is the solution of the quantum mechanical equation of the periodic potential field in crystal. Of course, it is different from the solution of the equation in vacuum, which causes the difference between the effective mass and the rest mass. In other words, in the quasi-classical method, the existence of the crystal periodic potential field is reflected in the effective mass m^*.

4.6   Effective Mass and Band Structure

The effective mass depends on the periodic potential field generated by atoms in crystal and should be related to the band structure. In order to explain the relationship between the effective mass m^* and band structure of crystal, we first look at the relationship between the mass m_0 and band structure of vacuum.

Perhaps the above statement is quite confusing: only a crystal with a periodic potential field can be described by band structure. There are no atoms in a vacuum, not like a crystal. Where does the band structure of vacuum come from?

There are indeed no atoms that make up the lattice in vacuum, and the potential energy is very small, which can be made zero. However, a zero field can also be regarded as a special case of the periodic potential field, so band structure of vacuum can also be discussed. Moreover, band structure does not necessarily belong to the periodic potential field only. It has been confirmed that in the amorphous solid, electrons also have a band structure.

Band structure of vacuum is nothing more than the relationship between the energy E of free electrons in vacuum and the momentum p (or k). If we consider not only the electrons in vacuum, but also the general "particles" with rest mass m, the relationship between energy E and momentum k is described in two cases: m is equal to zero, and m is not equal to zero.

As shown in Fig. 4.12a, when the mass m of the particle is not equal to 0 and the moving speed is much lower than the speed of light, the energy E is proportional to the square of the momentum k ($E = k^2/(2\,m)$). The classical electron (mass m_0) that does not consider the relativistic effect falls into this situation; see the energy band diagram in the parabolic shape shown in Fig. 4.12a. The energy E here is only the kinetic energy of the particle, and does not include the intrinsic energy mc^2 inside the particle, which is expressed by Einstein's mass-energy relationship. This intrinsic energy mc^2 is shown as parabolic minimum in Fig. 4.12a, where c represents the speed of light. For particles with mass equal to zero, such as photons, the relationship between energy and momentum is not parabolic, but linear. For photons: $E = ck$, as shown in Fig. 4.12b, since the speed of photon $v = c$.

As mentioned above, there are two different types of energy band diagrams for mass m (not equal to 0 or equal to 0). This question can also be rephrased: If you are given a band diagram of a certain shape, how do you estimate the mass of the particle? If you review Fig. 4.12a, b again, it is not difficult to get the answer to the question: if the energy band diagram is a conical line like Figure b, the particle mass

relationship between energy-momentum

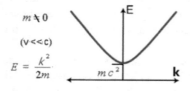
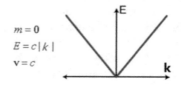

(a) parabolic for electron   (b) linear function for photon

Fig. 4.12 Band structure in vacuum (particle mass m)

is equal to 0; if the energy band diagram is a parabola, the particle mass not equal to 0. For the parabolic situation shown in Figure a, we can further obtain:

$$m = 1/(d^2E/dk^2)$$

The relationship between mass and band structure can be understood as follows: the mass of a particle m is a parameter in the band structure. The linear performance band curve corresponds to parameter $m = 0$. For the parabolic band curve, parameter m is the reciprocal of curvature, which can be expressed as second derivative of energy E versus momentum k.

The above statement is based on the vacuum energy–momentum relationship and it can be extended to band structure of crystals. In other words, from a certain band apex of band structure, its curvature (second derivative) can be calculated. The reciprocal of this curvature corresponds to a certain mass parameter m^*, which is the effective mass of electrons moving in the crystal. If the curvature does not exist at a certain vertex, it corresponds to the case where the effective mass m^* equals to 0, similar to the photon in vacuum described in Fig. 4.12b. Yet it is not completely equivalent to photons. At this time: $m^* = 0$, $E = vk$, so the velocity of Bloch electron $v = E/k$ is characterized by the slope of the line in band structure. The energy band diagram of graphene is cone-shaped, and the movement of electrons belongs to this type of situation where the effective mass is zero. In order to better understand the movement of electron in graphene, the meaning of effective mass will be discussed in depth below.

4.7   The Significance of Effective Mass

The effective mass m^*, to be more accurate, should refer to the effective "resting mass, or intrinsic mass" of the particle, just as m in the previous section refers to the resting mass of the particle. In general we just call it effective mass.

The relationship between the energy and momentum of the electrons in graphene is very special, described by a linear curve similar to Fig. 4.12b, that is, the effective mass of the electrons in graphene is zero. However, band structure of most semiconductors is approximately parabolic type as shown in Fig. 4.12a. Therefore, understanding movement of electrons with no-zero effective mass will help to understand movement of electron in graphene.

Generally speaking, the energy band curve E(k) of the crystal is not a strict parabola. The curvature of the curve is different from point to point, and the effective mass is inversely proportional to the curvature of the energy band. The effective mass is larger where the curvature is small, and vice versa. Therefore, in an energy band, the effective mass m^* is not a constant, but a function of k.

The introduction of effective mass has brought us great convenience in dealing with problems of moving electrons in crystal. The position of the moving electron is replaced by the center position of the Bloch wave packet so that it appears that Bloch electrons are like electrons in a vacuum, moving according to Newton's second law. The role of the lattice is only manifested in the effective mass. The electron velocity v and the effective mass m* are both determined by the shape of the band diagram of the electron in the crystal: the velocity v is proportional to the slope of the energy curve, and the reciprocal of the effective mass m* is equal to the second derivative of the energy curve to the wave vector k, namely the curvature of the curve.

Effective mass differs from the original concept of "mass" in another aspect. The mass in classical physics is an inherent property of matter; it does not change with the value of the wave vector k, and also a scalar quantity. Since the effective mass is defined as the "curvature" of energy band in the wave vector space, the curvature of each direction is different. This makes the effective mass not a scalar, but a tensor. Only under special conditions, when the band diagram has simple symmetry, the effective mass degenerates into a scalar. For simplicity, we only consider the scalar case with either a positive or negative value. Near the bottom of band, the effective mass is always positive $m^* > 0$; near the top of the band, the effective mass is always negative $m^* < 0$. This is because at the bottom and top of the energy band, E(k) takes the maximum value and the minimum value, respectively, and has a positive and negative bivalent derivative.

The effective mass m^* includes the effect of lattice forces so that Bloch electrons in the crystal are just a kind of "quasi-particles". The effective mass $m^* > 0$ indicates that the momentum obtained by the electron from the external field is greater than the momentum transferred to the crystal lattice; on the contrary, the effective mass $m^* < 0$ indicates that the momentum transferred to the crystal lattice by the electron is greater.

As shown in Fig. 4.13, the effective mass m^* can be greater than m or smaller than m, depending on the effect of the lattice force, assuming that the electrons initially

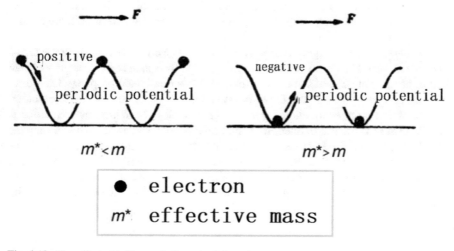

Fig. 4.13 The effect of lattice periodic potential on electrons

gather near the apex of the crystal potential field, as shown in Fig. 4.13a, when there is an external force, its effect is equivalent to pushing the electrons to "roll down" along the potential energy curve, so $m^* < m$. In another case, as shown in Fig. 4.13b, when the electrons gather in near the bottom of the potential energy curve, it is obvious that the effect of the lattice potential field is exactly the opposite of the external field force. The external field urges it to leave the lowest point, and the lattice potential field hinders it from leaving, which is expressed as $m^* > m$.

All in all, with the concept of effective mass and wave packets, the movement of Bloch electrons in the external field can be used to study the movement of Bloch's electrons in the outer field with Newton's second law. The wave packet expresses the idea of quantum theory, and the effective mass takes into account the effect of the periodic field of the unknown lattice. For example, if an external field F acts on an electron with an effective mass of m^*, the movement of the electron will follow Newton's second law:

$F = (m^*)a$, where a is the acceleration generated by the electron under the action of external field F.

Now, let's go back and discuss the case where the effective mass m^* is zero such as graphene. In the band diagram of the parabolic shape, the effective mass is related to the curvature. What is the curvature of the cone? Riemannian geometry has given the answer to this question for a long time. According to Riemannian geometry, a 2-dimensional cone embedded in a 3-dimensional space is a developable surface. The intrinsic curvature of the surface is zero everywhere, except for the curvature at the vertex which is infinite. Since the effective mass is the reciprocal of the curvature, the effective mass on the cone surface becomes infinite, but the effective mass at the apex is zero. The properties near the Fermi level in band diagram are the key to determining the characteristics of electrons in the material. The Fermi level in the graphene band diagram passes through the apex of the Dirac cone and is called the Dirac point. The behavior of electrons near this point is of our interest. Therefore, when we talk about the effective mass in graphene, we refer to the effective mass near the apex of the Dirac cone, and its value is zero.

The band structure of graphene is a cone shape. The effective mass near the top of the cone is zero. The Fermi velocity (about 1/300 of the speed of light) is greater than that of carriers in general semiconductors. This is a relativistic property. Therefore, moving of electrons near the Dirac point should be described by Dirac equation instead of Schrodinger equation. In quantum mechanics, the Schrodinger equation is used to deal with the wave function of electrons whose speed is much lower than the speed of light, and the Dirac equation is a quantum mechanical equation that takes into account the effect of relativity. In the next chapter, we will discuss the formation of energy bands, the Dirac cone of graphene, etc., and give a brief introduction to the Schrodinger equation and the Dirac theory combined with the effects of special relativity.

Chapter 5
Electrons Dance in Graphene

5.1 Band Structure of Graphene

We have seen previously that the E-k relation of graphene near the Fermi level is linear; it is a 2-dimensional projection of the diagram. Band structure of graphene is indeed very special, especially its structure near the 6 symmetrical K and K′ points in the first Brillouin zone as shown in Fig. 5.1. The band structure creates the extraordinary electrical and physical properties of graphene.

From the enlarged cone diagram on the right of Fig. 5.1, it can be seen that the conduction band and valence band in the pure graphene, as well as the Fermi level, intersect linearly at one point. This point is called the "Dirac point". Its 3-dimention shape is called "Dirac cones". In this chapter, we explore Dirac cone in more detail.

Although the real single-layer graphene was not manufactured until 2004, its band structure was studied by Canadian theoretical physicist P. R. Wallace as early as 1947.

The energy band of the Dirac cone shape is unique. The key physical properties of graphene are generated near the Dirac point. According to the discussion in the previous chapter, the band curve of most crystals is parabolic such that the movement of electrons is similar to the free electrons in vacuum, except that the mass of electrons should be replaced by an effective mass. The effective mass takes into account the influence of the lattice. It is like a person running in a crowd; another individual may block him or push him along the way. The same is true when the electron is in motion: the atoms in the crystal lattice may hinder it or help it. All these functions are combined and summed up with an effective mass.

Near the Dirac point of graphene, the conduction band and the valence band intersect linearly at one point, which shows that the electron's energy E and momentum k are linearly dependent, similar to a photon with zero-mass in vacuum. This shows that the effective mass of the electron has become zero. The effect of the graphene lattice atoms does not impede the electrons, but instead makes the electrons have no resting mass! The electrons can be transported smoothly and run at the maximum speed. This behavior of Dirac fermions with an effective mass of zero near $k = K$

© Guangxi Science & Technology Publishing House 2022
T. Zhang, *Graphene*, https://doi.org/10.1007/978-981-16-4589-1_5

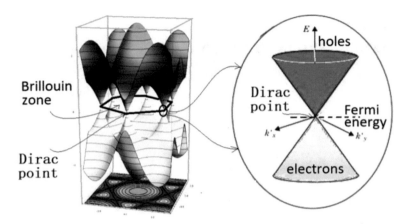

Fig. 5.1 Energy band of graphene

has been confirmed by graphene experiments. The movement of electrons cannot be described by the non-relativistic Schrödinger equation near by the Dirac point, rather by the Dirac equation of quantum electrodynamics.

How is the Dirac cone shape of the graphene calculated theoretically? Or we can ask more general question: how is the band structure of each crystal obtained? It is basically impossible to accurately solve the quantum equations according to the atomic-electron interaction in the crystal. This is why we use band theory as a valuable asset in solid-state physics. Band theory is originally based on approximations of electrons and periodic field.

Before explain Dirac cone of graphene, let us first qualitatively review approximation methods used to solve the electron motion of crystals in general. We will also give a simple introduction of Schrödinger equation, Dirac equation, and special relativity.

5.2 Approximation Methods

The movement of electrons in crystals has two extremes: free electrons or bound electrons, each correspond to continuous band diagrams and separate energy levels respectively. There are two main methods for calculating the band diagram of periodic potential: free electron approximation and tight-binding approximation.

If electrons are fully shared and move freely, their energy can be continuously taken to any value, just like free particles in a vacuum. The energy band diagram in vacuum is shown in Chap. 4. Figure 4.12a is a parabola for particles with non-zero mass and Fig. 4.12b is linear relationship for photons. On the contrary, if the electron is only bound to move around an isolated atom, the energy of the electron can only take discrete values, the band diagram is completely discontinuous and becomes discrete energy levels.

In most cases, electron movement in crystals is neither completely free electrons, nor is it bound by an isolated atom. The band diagram of most crystals is between "continuous band" and "discrete energy levels". The possibility of electron energy value is represented by both, continuous part (allowable band) and discontinuous part (forbidden band). Band structure of crystal has composed of alternate phases.

Both the allowable band and the forbidden band indicate that in the crystal have both common and bound electrons. We can start from the above two extreme cases, gradually calculate and approximate the true movement of electron in the crystal step by step, and get a more accurate band diagram theoretically. These lead to the two approximation methods.

5.2.1 Nearly-Free Electron Model

Assuming that the crystal potential is weak and the average kinetic energy of electrons is much larger than the crystal potential field, the behavior of crystal electrons is closer to that of free electrons in vacuum. The vacuum here must also meet the requirements of crystal translational symmetry in the problem. Therefore, the so-called free electron approximation starts from the band diagram of free electron of the lattice, the parabola described in Fig. 5.2a, and the parabola is regarded as the zero-order approximation of band diagram. Then, the periodic potential field of atoms is regarded as a perturbation, and its correction to the parabola is calculated and solved.

In the near free electron approximation, the perturbation of the periodic potential field causes discontinuities in the single parabolic band of the lattice, as shown in Fig. 5.2b (at $|k| = 2\pi/a$ and $|k| = \pi/a$), where a is the lattice constant. Therefore, the gap can be opened to form a band gap. In other words, due to the influence of the periodic

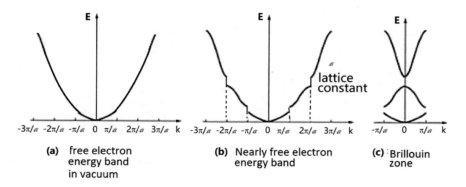

Fig. 5.2 Near free electron approximation

potential, electron waves of certain wavelengths are canceled with themselves due to the reflection of the lattice points, and cannot propagate in the lattice like the original free electrons. It is as if the electron interacts with the periodic potential field and is directly "hit" to the new momentum state, instead of following the original pure parabolic rule. The original continuous energy–momentum relationship becomes a discrete energy band, which is folded into the first Brillouin zone to become our common energy band diagram.

Nearly-free electron model is simple, and it can open the degeneracy to form a forbidden band near the boundary of Brillouin zone. Because only the electrons satisfying the boundary reflection conditions of the Brillouin zone can form a standing wave, some electrons corresponding to a specific energy are not allowed to exist, resulting in an energy gap. Nearly-free electron approximation is suitable for situations with good periodicity and strong interaction between atoms, such as metals.

5.2.2 Tight-Binding Approximation

This approximation is the opposite of nearly-free electron calculation. The method is suitable for weak periodicity or weak interaction between atoms. In the tight-binding approximation, the potential of a single atom is very strong, the crystal electrons are tightly bound around it, and the position of the electrons becomes discrete lattice points. These lattice points represent the actual positions of the atoms in the crystal, as shown in upper left picture of Fig. 5.3. In other words, the solution of the Schrödinger equation obtained by an isolated atomic potential well is regarded as the zero-order approximation, and the solution at this time should be the discrete energy levels that the electron may occupy. Starting from these discrete energy levels (Fig. 5.3, left), the interaction between each atom and its neighbors is treated as perturbation.

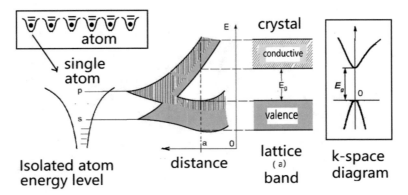

Fig. 5.3 Tight-binding approximation

The smaller the distance between an atom and its neighboring atoms, the greater the perturbation, to a certain extent, electrons can jump between different lattice points, and the jump probability corresponds to the "overlap" of the actual electron wave functions of different atoms. The wave function of an electron is represented by a linear combination of the wave functions of all atoms.

Starting from isolated atoms, gradually consider the interaction between atoms, which corresponds to the continuous reduction of the distance between atoms, the continuous splitting of energy levels, and the sharing of electrons by more atoms. When the distance between atoms is equal to the lattice constant, it represents the real situation of the crystal. At this time, the discrete energy levels are broadened into the corresponding energy bands: the conduction band and the valence band with the band gap Eg, as shown in the figure below. When there is large number of electrons the crystal is shared by all electrons in the crystal lattice, it is impossible to distinguish each energy level, so bands appear to be continuous. Transform from the lattice space to the wave vector space, and get common band diagram, which is the rightmost diagram in Fig. 5.3.

As mentioned earlier, the free electron approximation and the tight-binding approximation start from different physical considerations, use different perturbation factors, and achieve the same goal by different routes. In summary, the free electron approximation explains how the forbidden band is generated, and the tight-binding approximation explains the formation of the continuous band.

From a chemical point of view, it can also be said that the tight-binding approximation actually treats the solid as a huge "molecule". Isolated atoms have discrete energy levels. After two atoms form molecules, the electron wave functions overlap to form molecular orbitals. If you think of a solid as a huge molecule composed of atoms, because of the large number of atoms, the total number of orbitals is also large, and the energy levels formed are denser, which can be regarded as a continuous energy band.

5.3   Schrodinger Equation

The two types of equations in quantum physics are often mentioned in this book, but the specific form is not written. In this section we will first briefly introduce Schrodinger equations of quantum mechanics.

Schrödinger was an Austrian physicist. What was engraved on the marble bust of Schrödinger placed at the University of Vienna is the simplified Schrödinger equation:

$$i\hbar\dot{\psi} = H\psi$$

There are energy operators on both sides of the equation. If written in a more detailed differential equation form and swapped for left and right, it is:

$$-\frac{\hbar^2}{2m}\frac{\partial^2}{\partial x^2}\Psi(x,t) + V(x)\Psi(x,t) = i\hbar\frac{\partial}{\partial t}\Psi(x,t) \tag{5.1}$$

The Schrödinger equation looks complicated, yet it is simple to understand, it corresponds to the relationship between energy E and momentum p learned in classical mechanics and high school physics textbooks:

$$P^2/2m + V = E \tag{5.2}$$

The first term ($P^2/2m$) in the above formula (5.2) is the kinetic energy of the particle, the second term V is the potential energy of the particle, E is the total energy. In quantum mechanics, all physical quantities of classical mechanics are corresponding to a certain operator, so momentum p and energy E are replaced by their corresponding differential operators. The energy operator ($\partial/\partial t$) and the momentum operator ($\partial/\partial x$) are the (partial) derivatives with respect to time and space, respectively, which should multiply by ih. After replacing them with operators, the formula (5.2) is written as (5.1). If you consider time-independent systems (such as crystals), the equation can be written as:

$$H\psi = E\psi \tag{5.3}$$

The energy operator H, also called the Hamiltonian operator, occupies an important position in quantum mechanics. The energy operator H in the formula (5.3) is the sum of kinetic energy and potential energy. E is the eigenvalue of energy. Therefore, solving the Schrodinger equation becomes a problem of solving the characteristic equation of Hamiltonian. For a single-atom system, all the discrete energy eigenvalues E_i that are solved correspond to the energy level of the electron, which is the possible value of the energy E.

The energy operator H is not necessarily expressed as the classical form of kinetic energy plus potential energy in formulas (5.1) and (5.2). For example, if the particle is moving fast and special relativity needs to be considered, the relationship between the energy and momentum of the particle will be different.

5.4 Special Theory of Relativity

Einstein is a well-known scientist throughout the world, his most famous physics formula: $E = mc^2$.

This is called the "mass-energy relationship" in physics. In the formula, E is energy, m is the rest mass of the object, c is the speed of light. The meaning of this formula can be roughly understood as follows: Corresponding to an object with a mass of m, there is energy with a value of E ($= mc^2$), which is bound inside the object. In other words, in a certain sense, mass and energy can be considered equivalent and can be converted into each other. Because the speed of light c is a very large value, the value of energy E obtained from the mass-energy relationship is also very large.

How was the mass-energy relationship derived? It is not the content of Newtonian mechanics that we are familiar with in middle school physics, but comes from Einstein's special theory of relativity.

One of the differences between special relativity and classical Newtonian mechanics is the understanding of the concept of "time". In classical Newtonian mechanics, time is absolute, while in relativity, time and space are related to each other. This difference is briefly described below.

Consider the transformation formula when two coordinate systems move relative to each other. For example, when two coordinate systems move at relative speed u in the x direction, Newtonian mechanics considers t to be unchanged, and the transformation is described by Galilean transformation.

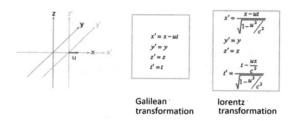

Galilean
transformation

lorentz
transformation

In relativity, the Galilean transformation replaced by Lorentz transformation, which can be derived from the two basic postulates: the equivalence of all inertial reference frames and the invariance of the speed of light. The first one called "principle of relativity", and second one means that the propagation speed of light in vacuum is a constant relative to any observer. According to the special theory of relativity, in a four-dimensional space–time following the Lorentz transformation, the energy E of a particle with a rest mass of m is:

$$E = \frac{mc^2}{\sqrt{1 - \left(\frac{u}{c}\right)^2}} \quad \text{or} \quad E^2 = \frac{m^2c^4}{1 - \left(\frac{u}{c}\right)^2} \tag{5.4}$$

$$\text{small u} \quad \underset{\text{back to Newton}}{\overset{u \to 0}{\Rightarrow}} \quad E = mc^2 + \frac{1}{2}mu^2 \tag{5.5}$$

The energy formula (5.4) evolves into (5.5) when the speed u is very small. The energy E of (5.5) contains two parts. The second term is the kinetic energy of the particle in Newtonian mechanics, and the first term can be regarded as the energy inside the particle. When the speed u = 0, the kinetic energy part is zero, and we get: $E = mc^2$, which is the well-known mass-energy relationship as described above.

5.5 Dirac Equation

From the introduction in Sect. 5.3, there is an interesting analogy between the Schrodinger equation and the Newton's equation. To put it simply, starting from the Newtonian energy–momentum (E-p) relationship (5.2), replacing E and p with their operator's $\partial/\partial t$ and $\partial/\partial x$ respectively (multiply ih), the Schrodinger Eq. (5.1) can be obtained formally.

When considering the special relativity, E-p relationship the particle should satisfy is described by (5.4). If the velocity u is relatively small, the fractional form can be altered slightly, and the velocity u can be expressed by the particle momentum p, where $u = p/m$, then, the formula (5.4) becomes:

$$P^2c^2 + m^2c^4 = E^2 \tag{5.6}$$

Combined with the special relativity, the E-p relationship is different from the Newtonian relationship (5.2), and the quantum mechanics equation obtained after replacing it with operator is of course also different. Following Schrodinger's method, Swedish physicist Klein and German physicist Gordon replaced E and p in Eq. (5.6) with corresponding differential operators, and obtained Klein-Gordon's equation:

$$\frac{1}{c^2}\frac{\partial^2}{\partial t^2}\psi - \nabla^2\psi + \frac{m^2c^2}{n^2}\psi = 0$$

The Klein–Gordon equation is unsatisfactory to use due to the fact that there is an E^2 term that makes the equation a second-order differential with respect to time. This is completely different from the Schrodinger equation, which is a first-order differential equation with respect to time.

Dirac solved this problem in an ingenious way and created relativistic quantum mechanics, for which he won the Nobel Prize in Physics in 1933.

Paul Dirac appreciates the beauty of mathematics all his life. He said: "God used beautiful mathematics to create this world." Dirac felt that the Klein–Gordon equation encountered the dilemma of E^2. How to get a first-order differential equation with respect to time that satisfies the theory of relativity? Dirac, who likes to solve mathematic problems, immediately thought of taking a square root on both sides of the E-p relation (5.7):

$$H_{\text{Dirac}} = \text{Sqrt}(p^2c^2 + m^2c^4) = E \tag{5.7}$$

Since p is an operator in expression $(p^2c^2 + m^2c^4)$, how to get the results when you take "square root" of operators? Dirac tried various approaches and finally was able to successfully obtain the operator after the square root:

$$H_{Dirac} = c\sum_i \alpha_i p_i + \beta mc^2$$

This is called the Dirac Hamiltonian. The final Dirac equation is:

$$(c\boldsymbol{\alpha} \cdot \hat{\boldsymbol{p}} + \beta mc^2)\psi - i\hbar\frac{\partial\psi}{\partial t} \tag{5.8}$$

Here:

$$\beta = \begin{pmatrix} 0 & 0 & 1 & 0 \\ 0 & 0 & 0 & 1 \\ 1 & 0 & 0 & 0 \\ 0 & 1 & 0 & 0 \end{pmatrix}, \alpha_1 = \begin{pmatrix} 0 & -1 & 0 & 0 \\ -1 & 0 & 0 & 0 \\ 0 & 0 & 0 & 1 \\ 0 & 0 & 1 & 0 \end{pmatrix}$$

$$\alpha_2 = \begin{pmatrix} 0 & i & 0 & 0 \\ -i & 0 & 0 & 0 \\ 0 & 0 & 0 & -i \\ 0 & 0 & i & 0 \end{pmatrix}, \alpha_3 = \begin{pmatrix} -1 & 0 & 0 & 0 \\ 0 & 1 & 0 & 0 \\ 0 & 0 & 1 & 0 \\ 0 & 0 & 0 & -1 \end{pmatrix} \tag{5.9}$$

They satisfy:

$$(\alpha_i)^2 = \beta^2 = I_4$$
$$[\alpha_i, \alpha_j]_+ = 0$$
$$[\alpha_i, \beta]_+ = 0 \tag{5.10}$$

I_4 is an unit matrix with 4 rows and 4 columns. The above 4×4 matrices can actually be formed by four 2×2 Pauli matrices for spin:

$$\sigma^0 = \begin{pmatrix} 1 & 0 \\ 0 & 1 \end{pmatrix}, \sigma^1 = \begin{pmatrix} 0 & 1 \\ 1 & 0 \end{pmatrix},$$

$$\sigma^2 = \begin{pmatrix} 0 & -i \\ i & 0 \end{pmatrix}, \sigma^3 = \begin{pmatrix} 1 & 0 \\ 0 & -1 \end{pmatrix},$$

The Dirac equation not only combines the theory of special relativity with quantum mechanics, but also automatically integrates the mysterious concept of "electron spin" into the equation.

5.6 Dirac Cone

Although the Dirac cone is not exclusive to graphene, it is still a relatively rare and unique band structure. Its energy band is conical up and down at the Fermi energy level separating filled and unfilled electrons. So far, hundreds of two-dimensional materials have been discovered, but only graphene, silicene, germanene, and a small number of other materials are considered to have Dirac cones. So far only the Dirac

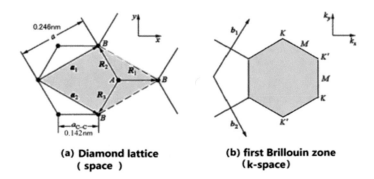

**(a) Diamond lattice
(space)**

**(b) first Brillouin zone
(k-space)**

Fig. 5.4 Graphene structure

cone in graphene is confirmed by experiments using the tight-binding approximation
method.

Figure 5.4a shows the unit cell of graphene lattice. It is a rhombus (the shaded
part in the figure) consisting of two unequal atoms A and B. Its lattice constant a
= 0.246 nm and the interatomic distance a_{c-c} = 0.142 nm. The reciprocal grid is
a hexagonal grid rotated 30° relative to the normal grid. The first Brillouin zone is
represented by a shaded hexagon in Fig. 5.4b. Corresponding to the unequal A and
B atoms, there are two unequal vertices K and K' in the first Brillouin zone.

Among the four valence electrons of a carbon atom, only the π-bonded electron
in the $2p_z$ state contributes to the common lattice. Because each unit cell has two
atoms A and B, there are also two π electrons, which occupy two adjacent energy
levels: $(2p_z)_1$ and $(2p_z)_2$, respectively. Finally, these two energy levels are expanded
into π band (filled valence band) and π* band (empty band, or conduction band) in
single-layer graphene. At the 6 vertices of the first Brillouin zone, the π band and the
π* band are degenerate, Fermi surface is reduced to 6 intersections. Using the tight-
binding approximation, start with wave function of two unequal AB carbon atoms
in the primitive cell, then add in the interaction between each atom and the three
nearest neighbor atoms to solve for the equation. For example, in the gray diamond
in Fig. 5.4a, consider the interaction between an A-atom and its nearest 3 B-atoms
(the distances from A to B in the figure are R_1, R_2, and R_3). Total wave function is
a linear superposition of each wave function, and finally we obtain the expression
of the corresponding band. For the detailed calculation process, please refer to Ref.
[17].

Because each unit cell of the graphene lattice has two atoms and two π electrons,
the final E-p relationship can be obtained by tight-binding approximation. Hamiltonian of graphene near the vertex K' of the first Brillouin zone can be expressed
as:

$$h(K' + q) = \overline{h}v_F \vec{q} \cdot \vec{\sigma} \tag{5.11}$$

Similar equations can be derived for other vertices (such as K). Compare the formula (5.11) with the Hamiltonian in the Dirac equation introduced in the previous section:

$$H_{Dirac} = c \sum_i \alpha_i p_i + \beta m c^2 \qquad (5.12)$$

The following conclusions can be drawn:

I. Formula (5.11) is the two-dimensional correspondence of (5.12), the matrix α_i in the four-dimensional Dirac Hamiltonian is replaced by a two-dimensional Pauli matrix;

II. In formula (5.11), q is a momentum operator, which corresponds to pi in (5.12);

III. There is no $\beta m c^2$ term in the formula (5.11), which means that the corresponding mass m = 0;

In the formula (5.11), the energy h(K' + q) has a linear relationship with q. When q = 0, the energy h(K' + q) is also equal to zero. This point is similar to (5.12): H_{Dirac} is linear with pi. Although in (5.11) and (5.12), the E is linear with p, the slope of the straight line is different. The place where the speed of light c in (5.12) is replaced by the Fermi speed v_F in (5.11).

Because the Hamiltonian of band structure of graphene as described in (5.11) is similar to the Hamiltonian of the Dirac equation describing relativistic particles, it is called Dirac cone. When the p is equal to 0, the energy E is also equal to 0 (k = 0), and the corresponding point is the Dirac point. The electron described by this energy band is a particle with an effective mass of zero, which behaves like a photon but has a different speed (v_F).

Since the spin of electrons is automatically included in the Dirac equation, the properties of electrons near the six Dirac points in the first Brillouin zone of graphene are important and helpful for understanding the quantum spin Hall-Effect, which will be introduced in the next chapter.

Chapter 6
Quantum Topology

6.1 Rubber-Sheet Geometry

Topology and geometry both are topics of mathematics to study space, but in very different ways. Topology is interested in how to make the connection between points and not so in the distance between them. For example, there are two geometric figures, we can stretch and deform them like clay, but they cannot be torn and pasted. If we can convert them to each other through the above methods, we would say these "two geometry have the same topology". Therefore, topology is also commonly referred to as "geometry on rubber film". The most typical example of different geometric shapes but "same topology" is what people often call "bagels and coffee cups". Here we give more examples intuitively to introduce several concepts of topology.

The applications of topology in theoretical physics are mainly condensed matter physics, quantum field theory and cosmology. Graphene is related to condensed matter physics, in which the introduction of concepts of topology is accompanied by the discovery of the quantum Hall effect and topological insulators.

6.1.1 Manifold and Topology

Manifold is a generalization of Euclidean space. Euclidean space is flat space such as straight lines and planes that we are familiar with. If you expand this concept a little bit, as long as every infinitely shrinking part of the space looks the same as a partial Euclidean space, it can be called a manifold. For example, connecting a wire to form a circle like "0" is an example of 1-dimensional manifold, but if it is connected to form a figure like "8", it is not a 1-manifold, because near the intersection of "8", it cannot be locally equivalent to a straight line.

Sphere, torus, doughnut, Mobius tape, and Klein bottle are all examples of 2-dimensional manifolds. The small parts near each point of them look similar to a

© Guangxi Science & Technology Publishing House 2022
T. Zhang, *Graphene*, https://doi.org/10.1007/978-981-16-4589-1_6

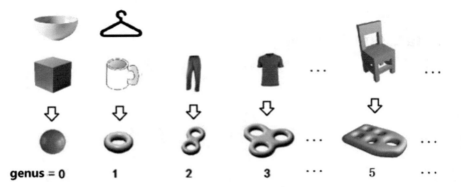

Fig. 6.1 Different types of topological manifolds represented by different genus

plane, but their overall topology is quite different. Therefore, examples listed above are manifolds with different topology.

6.1.2   Genus

The two-dimensional manifold is the most intuitive and interesting. Among them, manifolds such as spherical and doughnut surfaces are "limited, unbounded, and directional". They have been studied intensively and can be described and classified by "genus". For real closed surfaces, intuitively, genus is the number of holes on the surface, as shown in Fig. 6.1. When the genus number is equal to 1, it corresponds to the topological manifold represented by the doughnut or coffee cup.

6.1.3   Topological Invariants

Topological research cares about the connection between points, a certain internal property between two isomorphic topological spaces. This common property can be described by topological invariants. If a piece of dough is shape as a donut, then we can knead it gradually to a coffee cup, if we keep the shape of dough with 1 hole. In the above example "1 hole" is a "topological invariant" in the process of deformation. It means that we need to keep certain intrinsic properties of dough unchanged. Intuitively speaking, the "hole" that should maintain the dough always exists, and no new hole can be created. In other words, keep the number of "genus" of this topological manifold equal to 1. Therefore, loosely speaking, the number of genus is the typical topological invariant of 2-dimensional manifold in Fig. 6.1.

The "Chern number" in the next fiber bundle example is also a topological invariant.

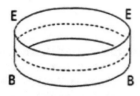
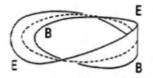

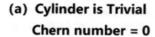

(a) Cylinder is Trivial   **(b) Mobius belt non-Trivial**
Chern number = 0   **Chern number = 1**

Fig. 6.2   Fiber bundle

6.1.4   Fiber Bundles and Chern Class

The fiber bundle can be regarded as a generalization of the product space. There are many examples of simple product spaces. For example, a two-dimensional plane XY can be regarded as the product of two one-dimensional spaces X and Y; a cylindrical surface can be regarded as a product of a circle and a one-dimensional straight line space.

The fiber bundle is the product of two topological spaces, the base space and the tangent space (fiber). The plane can be regarded as a cluster where X is the base and Y is the tangent space; the cylindrical surface can be regarded as a fiber cluster where the circle is the base and the one-dimensional straight line is the tangent space. It's just that the plane and the cylinder are both trivial fiber bundles, which means that the multiplying method of two spaces is the same at every point in base space. If they are not the same, it may be a non-trivial fiber bundle, such as the Mobius belt, as shown in Fig. 6.2.

Use part of human as a simple and intuitive image of the fiber bundle: human head is the base, hair is the fiber, and the head full of hair is a fiber bundle.

As mentioned above, fiber bundles can be divided into trivial and non-trivial. Topological property of fiber bundles can be classified by "Chern class" named by the mathematician Shiing-Shen Chern. An invariant—"Chern number" is zero or non-zero, can be used to characterize the difference in topological properties between the cylindrical surface and the Mobius belt fiber bundle in Fig. 6.2. Chern number can intuitively understand how many times the fiber rotates around the base space when the point of the base space changes. It can be seen from Fig. 6.2 that compared to a straight cylindrical surface, when the base space parameters change once, the "fibers" on the Mobius belt "twist" around the base space.

6.2 Classic Hall Effect

How is topology related to graphene? We have to start with the Hall effect. There are many kinds of Hall Effect, we will start with the introduction of the classic ancestor.

The classic Hall Effect was discovered by Edwin Herbert Hall (1855–1938) in 1879. It is said that the energized conductor in the magnetic field will be subjected to force. In present days it is commonly taught in high school physics classes. What is the reason for this force? Lorentz force can be used to easily explain the phenomena. In Hall's era more than 140 years ago, it was a daunting task to draw this conclusion. At that time, Maxwell had just established the electromagnetic theory, people were still ignorant of the atomic structure, and the electrons had not yet been discovered. Even the founder of electromagnetic theory like Maxwell has not taken a good look at this detail in his theory. He believes that the force received by a conductor with current in a magnetic field is a mechanical force on the conductor, but not the electromagnetic force acting on the current.

The nameless boy Hall did not believe in authority. He questioned Maxwell's conclusions and carried out rigorous and careful experiments. After countless failures, Hall was finally successfully. The Hall Effect was then named after him. He found that the current passing through the gold foil was indeed affected by the magnetic field, and thus generated a voltage perpendicular to both the current and the magnetic field. This voltage was called Hall voltage by later physicist, as shown in Fig. 6.3a.

Before the discovery of electrons, it was difficult to truly understand the essence of the classic Hall Effect. This is why even Maxwell believed that "only mechanical force acts on the conductor, and no force acts on the current." People here are talking about "current" instead of "electrons" because at that time, no one knew what electrons were, and it was completely unclear that the conduction mechanism of metals was caused by the movement of free electrons. Thomson discovered electrons in 1897, about 20 years after the discovery of the Hall Effect.

After the discovery of electrons, with the in-depth research on the conduction mechanism of metals and semiconductor materials, the understanding of the Hall Effect has been extended. In the framework of Maxwell's electromagnetic theory,

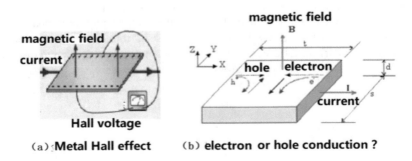

(a) Metal Hall effect   (b) electron or hole conduction ?

Fig. 6.3 Hall effect

the classic Hall Effect can be easily explained. The charge moving in the magnetic field will be subjected to the Lorentz force, causing the free electrons moving in the metal to produce an additional lateral movement, which piles up in a direction perpendicular to the original current to form a lateral voltage. This voltage is the Hall potential, which prevents the continued accumulation of charges, and will finally play a role in balancing the Lorentz force.

The discovery of electrons gave us a clearer picture of the atom, and concluded that metals can form electric currents because of the movement of electric charges. The overall metal itself is neutral, and the charge includes negatively charged electrons and positively charged protons in the nucleus. So, is the current formed because of the flow of electrons, or is it formed by the movement of atomic nuclei? Most people will come to the conclusion: Of course, it is electrons with a much smaller mass moving. However, this conclusion comes from your intuitive guess. How do you use experiments to prove this?

The Hall Effect provides us with such an experimental proof. Because the Hall Effect is caused by the movement of electric charges, is it positive or negative? The direction of the Lorentz force they receive is different, resulting in a different direction of the Hall potential.

Figure 6.3a shows the Hall Effect caused by the movement of free electrons in the metal. The relationship among the three directions of magnetic field, current, and Hall voltage is shown in Fig. 6.3a. If in semiconductor materials, the moving charge (or the carrier), is not necessarily an electron, it may also be a positively charged "hole". The direction of Hall potential generated by "hole" will be different. Therefore, we can use the Hall Effect to study the carriers in the semiconductor and determine the type of carriers in the doped semiconductor material. Is it a hole or an electron? The concentration of carriers can also be further measured.

Suppose that in a certain semiconductor material, the current is in X direction and the magnetic field is applied in Z direction, then what direction will Hall potential be? The answer depends on which type of majority carrier is in the material. Is it positive or negative? Let us discuss these two situations separately. As shown in Fig. 6.3b, if the current in the X direction (to the right in the figure) is caused by the moving electrons, the electrons move to the left. At this time, the force acting on the electrons is in the Y direction. In another case, if the current is caused by the movement of a positively charged hole, the direction of the hole movement is the same as the current, that is, it moves to the right. At this time, the Lorentz force acting on the hole is also at Y Direction. In other words, no matter whether the conduction mechanism is holes or electrons, the direction of Lorentz force is the same. Because electrons and holes have opposite charges and opposite moving directions, these two "opposites" cancel each other, resulting in the "same" direction of the final lateral movement.

However, the same direction of lateral movement does not mean the same direction of Hall voltage. Because of the different signs of the charges carried by carriers, that these two conduction mechanisms will form Hall voltages with opposite polarities. Therefore, we can determine the type of carriers in the material according to the polarity of the Hall voltage in the experiment.

Using the Lorentz force to explain the Hall Effect, it can be deduced that the Hall resistance is proportional to the magnetic field and inversely proportional to the carrier density in the conductor. In addition to studying the types of carriers in materials, the classical Hall Effect can also be used to measure carrier concentration or make magnetic sensors. Such Hall devices are used to detect magnetic fields and their changes, and have been widely used in various industrial applications related to magnetic fields.

After Hall discovered the conventional Hall Effect in non-ferromagnetic materials, in 1880, he discovered Anomalous Hall Effect in ferromagnetic metal materials. It means that when there is no external magnetic field, a transverse voltage will also be generated in the ferromagnetic materials with current. This phenomenon is confusing, because the Hall voltage in the metal is interpreted as the Lorentz force on the electrons. Since there is no external magnetic field, there is no Lorentz force, and the concept of Lorentz force cannot be used to explain the abnormality. Therefore, there is no unified theoretical explanation for the anomalous Hall Effect. It is generally believed that it is completely different from the normal Hall Effect in essence and cannot be explained by classical electromagnetic theory alone, but needs to be combined with the concepts of spin and orbit interaction in quantum theory.

So far, it has been more than 140 years since the discovery of Hall Effect. During the period, especially after the discovery of the quantum Hall effect in the 1980s, more family members of Hall Effect have been discovered, which has become a hot topic in condensed matter physics.

The Hall voltage is also often called the lateral voltage to distinguish it from the driving voltage along the current direction. The transverse voltage is directly proportional to the magnetic field vector B and the current intensity I, and inversely proportional to the thickness d of the metal plate.

According to the ratio of the horizontal voltage to the vertical current I, we can define a horizontal Hall resistance ρ_{xy}. This resistance should be proportional to magnetic field B. In classical Hall Effect, the relationship between ρ_{xy} and B is a straight line that slopes upward. The general longitudinal resistance ρ_{xx} should be a horizontal line that has nothing to do with the magnetic field, as shown in Fig. 6.4a.

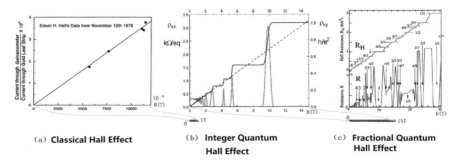

(a) **Classical Hall Effect**   (b) **Integer Quantum Hall Effect**   (c) **Fractional Quantum Hall Effect**

Fig. 6.4 Hall effect family

As mentioned earlier, Hall used gold foil for experiments, and the Hall Effect of metal was observed. If you use semiconductor wafers with different doping conditions, you will get different results. Physicists tried to use different materials, different thicknesses, in different environments, using different temperatures and magnetic fields, to study the Hall Effect. After experimenting for 100 years, they finally tried out a completely different alternative "Hall effect."

6.3   Quantum Hall Effect

The quantum Hall effect was discovered by the German physicist von Klitzing (von Klitzing, 1943–) in 1980, and won the Nobel Prize in Physics in 1985.

Comparing Fig. 6.4a, b it is easy to see the difference between the quantum Hall Effect and the classical Hall effect. The linear relationship between the Hall resistance ρ_{xy} and the magnetic field B in the classic effect is replaced by the more complicated curve in Fig. 6.4b. One by one "platforms" appeared in the transverse Hall resistance curve in Fig. 6.4b. The performance of the longitudinal resistance ρ_{xx} is quite different from the original classical one. The longitudinal resistance in the classical Hall effect, the ordinary resistance defined as the ratio of voltage to current in the usual sense is a constant, while in the quantum Hall effect, there was a abrupt change up and down.

It can also be seen from Fig. 6.4b that these platforms do not appear randomly. They appear at a certain value and correspond to a series of integer values. Moreover, values of the horizontal Hall resistance platform, which is the expression h/e^2, do not depend on materials and conditions used for experiment, but only determined by two basic physical constants: h and e. Here "h" is Planck's constant representing quantum effects, and "e" is electric charge representing electrons. This expression (h/e^2) happens to be the dimension of resistance; its value is approximately 25,813 Ω. The height of the resistance platform appears at the reciprocal of some integer multiple of this value, which is represented by n = 1, 2, 3... in Fig. 6.4b. The height of the highest one (n = 1) should be 1 (multiplied by h/e^2); the second is 1/2 (multiplied by h/e^2); the third is 1/3; then, 1/4, 1/5, (all need to be multiplied by h/e^2). In other words, the value of the Hall resistance platform is equal to (h/e^2) divided by an integer n. Each platform corresponds to an integer n. Because the quantum plateau value is associated with an integer, this phenomenon is called the integer quantum Hall effect.

In the quantum world, when the magnetic field increases continuously, the change of Hall resistance ρ_{xy} is not continuous. After it increases to a certain value, Hall resistance stops. Only when the magnetic field continues to increase to another value, the Hall resistance value suddenly jumps to a new value. As this continues, the platform is getting wider and higher, and the jump is getting higher and higher... The jump-like change of Hall resistance is exactly the "quantum" feature often mentioned by physicists, and it can only be explained by using quantum theory. This phenomenon excites physicists: In the seemingly boring experimental data, there is such a beautiful and moving quantum rhythm!

Observe the change in longitudinal resistance ρ_{xx}. In the classic case, the longitudinal resistance is parallel to the B axis, it has a fixed value and does not change with the magnetic field. This is consistent with our common sense, and the usual resistance should be independent of the magnetic field. It is quite different from quantum perspective. The curve of ρ_{xx} is not random and obeys a certain law: Whenever the Hall resistance ρ_{xy} appears on the platform, the resistance value ρ_{xx} will suddenly decrease to zero. This feature is also quite amazing, because the resistance of 0 means that the current can pass through the conductor unimpeded.

Readers may ask: Why did von Klitzing observe the Hall resistance platform in 1980, but the Hall Effect observed a hundred years ago is just a straight line. It is due to the development and progress of experimental technology in the past 100 years. The classic Hall Effect uses gold foil. The thinnest gold foil obtainable is still far away from the single-layer atomic graphene material. Therefore, the observed phenomenon can only be regarded as the Hall Effect in 3-dimensional metal. The most important difference is the experimental conditions: von Klitzing's laboratory can observe the quantum Hall effect, the most critical conditions are "deep low temperature and strong magnetic field." Von Klitzing's quantum Hall Effect is obtained at absolute 1.5 K degrees ($- 271$ °C) and a magnetic field as high as 19.8 T. The classical Hall Effect was originally observed at room temperature with a magnetic field of about one Tesla.

The difference in experimental conditions is also reflected in the abscissa of Fig. 6.4a, b. The abscissa marks the size of the magnetic field during the experiment. Figure 6.4a is the original data of Hall that year, and Fig. 6.4b is the data of von Klitzing. The magnetic field range of the entire abscissa of Fig. 6.4a is about 1 T, which is just a small section at the beginning of Fig. b. If we carefully observe this small section of the curve in Figure b, we will find that in the experimental results of that section: ρ_{xy} is linear and ρ_{xx} is a constant, which is completely consistent with the results of the classical Hall Effect.

The material used by von Klitzing is a thin "2-dimensional electron gas" inversion layer formed in the semiconductor material MOSFET, usually only a few nanometers thick. In this thin layer, electrons are completely trapped in the Z direction perpendicular to the sheet, but they can move freely in (X, Y) in the sheet. This structure exhibits many peculiarities under deep low temperature and strong magnetic field. Quantum Hall effect is one of them.

In 1982, American physicist Daniel C. Tsui and Horst L. Störmer of Bell Laboratories in New Jersey (USA) studied 2-dimenional electron gas under a high magnetic field (above 20 T), using a material with a higher carrier density (HEMT structure). They discovered fractional quantum Hall effect by obtaining finer steps than the integer quantum Hall effect (IQHE) curve. The Hall resistance is not only quantized, but also the constant h/e^2 divided by a certain fraction. The first observed fractional state is the 1/3 state, and then nearly 100 fractional states are observed.

Figure 6.4c is the results of fractional quantum Hall effect. As can be seen from the figure, in addition to integer platforms, there are many fractional platforms, so it is called the fractional quantum Hall effect. Physicists refer to these two Hall effects collectively as the quantum Hall effect.

6.4   Hall Effect in Graphene

Both integer and fractional quantum Hall effects are observed in 2-dimensional electron gas, indicating that the quantum Hall effect has a special favor of "2-dimensional" structures. So, now that we have real 2-dimensional crystals like graphene, what kind of wonderful Hall dance will electrons perform on it?

When Geim separated graphene from graphite for the first time, he could not wait to experimentally verify the integer quantum Hall effect of graphene. He found that the quantum Hall effect in graphene is different from the standard quantum Hall effect [18], as shown in Fig. 6.5. Also in 2005, another experimental team observed the fractional quantum Hall effect of graphene [19].

It can be seen from Fig. 6.5 that in the integer quantum Hall effect of graphene, the Hall resistance ρ_{xy} (or conductance σ_{xy}) does not have a $n = 0$ platform, and its quantization conditions can be described as:

$$\sigma_{xy} = \pm g_5(n + 1/2)e^2/h = \pm v\, e^2/h, \quad \text{where} \quad v = g_5(n + 1/2)$$

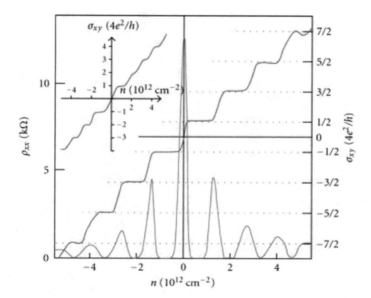

Fig. 6.5 The integer Hall effect of graphene (https://www.researchgate.net/publication/748 7268_Two-Dimensional_Gas_of_Massless_Dirac_Fermions_in_Graphene)

The platform corresponding to Landau fill factor $\nu = \pm 2, \pm 6, \pm 10, \pm 14 \ldots$, which can be seen clearly in Fig. 6.5. Compared with the general integer quantum Hall Effect, the quantization condition of graphene has a shift of 1/2. The reason is the relativistic effect of graphene at the Dirac point and the degenerate states of electrons and holes.

The small plot in the upper left corner of Fig. 6.5 shows the integer quantum Hall effect of double-layer graphene.

The graphene quantum Hall effect has a very special advantage: it can happen at room temperature! Most of the Hall Effect can only be observed at low temperatures (below 4.2 K). Since the electrons near the Dirac point of graphene are massless relativistic fermions, the graphene carriers have extremely high mobility. This property does not change much from low temperature to room temperature, so that at room temperature The quantum Hall effect of graphene is still observed (Fig. 6.6) [20].

The quantum Hall effect at room temperature is very important for the manufacture of electronic devices, because it does not require the use of liquid helium for refrigeration and is easier to promote and apply. This advantage makes graphene the first choice for next-generation electronic devices.

The Hall Effect of graphene is indeed very special, but so far, it still seems to have nothing to do with topology. Don't worry, everyone, we have to go back explore and explain the quantum Hall effect a little bit to figure out how those resistance platforms are produced.

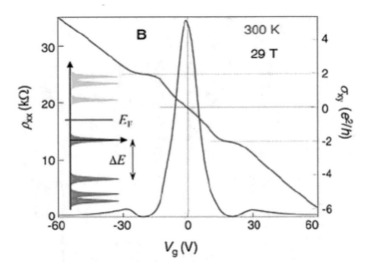

Fig. 6.6 Quantum Hall effect of graphene at room temperature (https://calame.unibas.ch/wp-content/uploads/teaching/mces/2017/students/Fabian_Mueller-RT_QHE_in_Graphene-.pdf)

6.5 Laughlin State

We are going to briefly cover the quantum Hall effect in this section.

In order to better understand the quantum Hall effect, it is good to revisit the classical motion of electrons in a magnetic field that we have learned from high school physics.

For a classical 2-dimensional electron moving in a uniform magnetic field, the magnetic force (Lorentz force) follows the right-hand rule and should be perpendicular to the direction of motion (Fig. 6.7a). Since the magnetic field does not work on the electrons, the kinetic energy and speed of an electron remain constant. This means that the electrons will maintain a circular motion in a gyration, and the gyration radius is related to its initial momentum. If we do not consider the linear velocity of electron in detail and only use the angular frequency ω_c of its rotation to characterize the motion, we can temporarily hide this initial momentum. Lets only consider the cyclotron angular frequency, which is just related to the charge-to-mass ratio of electron and the magnetic field. In other words, classical electrons in a magnetic field dance in a classical way whose gyration rate increases as the magnetic field B increases, as shown in Fig. 6.7b.

If there is electric field acting on 2-dimensional plane with moving electrons, then electrons will move under Coulomb force of the electric field and will have circular motion. This is the theoretical basis for explaining the classic Hall Effect.

As shown in the left figure of Fig. 6.7b, if the value of the magnetic field B is relatively small, electrons have reached the boundary of the metal sheet before "circling", they will accumulate at the boundary to form a Hall potential, producing the classic Hall Effect.

When the magnetic field increases, the angular frequency of electron increases, the electron turns a smaller circle, as shown in the diagram in Fig. 6.7b. This is when electrons begin to move in a circular pattern, and begin to produce an integer quantum Hall effect.

Characteristic of quantum Hall effect is the steps on the Hall resistance graph, each step represents a discontinuity. This 'discontinuity' and 'one piece' are the basic characteristics of quantum in physics. The integer quantum Hall effect (IQHE) can

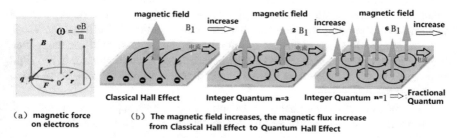

Fig. 6.7 The cyclotron dance of electrons in a magnetic field

be roughly explained by the concepts of energy band theory and Fermi level that we have introduced.

The electronic circular movement is a classic analogy. From a quantum point of view, are there different types of circular movement?

First, although the angular frequency ω_c of the classical electron is fixed under a certain magnetic field, their energy can continuously change with the cyclotron radius. However, the energy of the electrons dancing in quantum sense cannot change continuously, the interval of change can only be an integer multiple of its angular frequency ω_c.

Each electron is moving individually; it does not always contribute to the current. Only the electrons near Fermi level contribute to the current output. The parameter n, which is the characteristic of the integer resistance platform, can be understood as the ratio of the number of electrons N to the number of magnetic fluxes N_ϕ in 2-dimensional electronic system, that is, $n = N/N_\phi$.

Because the electrons are moving inside 2-dimensional material, the result of the gyration is that the current in the middle of the material is zero, the currents of two adjacent gyros cancel each other. The electrons on the boundary cannot form a complete cycle and eventually only move in one direction. Therefore, there is only marginal current in the quantum Hall effect.

For the quantum Hall Effect, there are two aspects of quantization that need to be considered: one is the quantization of electronic motion, which results in the Landau level. In addition, the quantization of the magnetic field must also be considered. The magnetic field generates magnetic flux in the system. When the magnetic field interacts with electrons, this magnetic flux should also be quantized. In other words, total magnetic flux can be divided into quanta of magnetic flux. Value of each flux quanta is equal to h/e. Here h is Planck's constant and e is the electronic charge. Although the intensity of the magnetic field seems to change continuously, for each electron, only when the magnetic flux that affects its motion becomes an integer multiple flux quanta, the wave function of the electron can form a stable standing-wave quantum state.

In a 2-dimensional system with a limited area, suppose the total number of electrons is N and the number of magnetic flux quanta is N_ϕ, their ratio, N/N_ϕ, corresponds to the integer n in the integer quantum Hall effect.

For example, in the middle picture of Fig. 6.7b, there are 6 electrons and 2 flux quanta ($N = 6$, $N_\phi = 2$), which is equivalent to sharing 1 flux quanta for every 3 electrons, corresponding to an integer platform of the quantum Hall effect $n = 6/2 = 3$.

If the magnetic field increases, the number of flux quanta will increase, then one magnetic flux quanta is shared by fewer electrons, n will decrease. In the right picture of Fig. 6.7b, 6 electrons share 6 magnetic flux particles ($N = 6$, $N_\phi = 6$), so $n = 6/6 = 1$.

What happens if the magnetic field continues to increase? That is to say, each electron will be divided into more flux quanta than n = 1. At this time, N_ϕ is greater than N so that $n = N/N_\phi$ becomes a fraction. It should be possible to get a value of n smaller than 1, 1/2, 1/3, etc. to get the fractional Hall Effect!

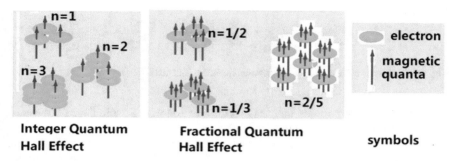

Fig. 6.8 "Candied haws" model of quantum Hall effect

In fact, even if the magnetic field is smaller than $n = 1$, it is possible to observe the fractional quantum Hall Effect. In that case, the ratio N/N_ϕ of the number of electrons N to the number of magnetic fluxes N_ϕ is a fraction greater than 1. Therefore, some people use the image of Fig. 6.8 to describe the distribution of the number of electrons and magnetic flux quanta for quantum Hall effect.

As shown in Fig. 6.8a, an electron is represented as a green circle in the figure, and the magnetic flux passing through the electron is represented by a blue arrow. It can be seen from Figure a, that the number of electrons passed by each magnetic flux quanta in IQHE is equal to the integer n.

When $n = 1$, a flux quanta passes through an electron, when $n = 2$, one flux quanta passes through two electrons and so on.

Now let us look at the fractional quantum Hall effect. The fractional plateau in the Hall Effect is observed for the first time when the total number of electrons remains unchanged and the magnetic field increases. After passing the $n = 1$ platform, if the magnetic field continues to increase, the number of flux quanta will also increase. Too many sticks and insufficient green circles, that is, the number of magnetic fluxes is too much and the number of electrons is not enough to distribute, so several magnetic quanta will share. The situation of an electron is shown in Fig. 6.8b. If two fluxes pass through an electron together, the corresponding integer n in IQHE becomes a fraction: $n = 1/2$; if three fluxes pass through an electron, $n = 1/3$. There are more complicated situations. For example, if five magnetic quanta pass through two electrons, then: $n = 2/5$.

The above model gives an intuitive image of the quantum Hall effect, but it is not a complete physical theory. The person who really gave a reasonable physical explanation for the fractional quantum Hall effect was R. B. Laughlin, who shared the 1998 Nobel Prize in Physics with the discoverers C. Tsui and L. Störmer.

The explanation of the integer Hall effect is based on the single electron approximation in solid theory, electrons move in the periodic potential field of lattice atoms. In other words, the single-electron approximation estimates the extremely complex many-body problem to an electronic problem to study, without considering the interaction between electrons. If you use the analogy of the electronic circular dance, the

Fig. 6.9 Topology corresponding to fractional quantum Hall state

single electron approximation means that the electron does a "solo dance", and each electron just dances its own circle independently.

However, the fractional quantum Hall effect is obtained at a lower temperature and a stronger magnetic field. Under this condition, the correlation between electrons cannot be ignored, on the contrary, this correlation plays a key role in determining the appearance of the fractional platform phenomenon in FQHE. This is like all the electrons dancing together in a collective style. In addition to its own solo dance, each electron also dances with every other electron. Therefore, the pattern of dance steps will be much more complicated.

We can also use the analogy of electronic dance to understand the candied haws model corresponding to the various quantum Hall effects in Fig. 6.8. For example, the integer Hall effect in Fig. 6.8a shows the "solo" mode of electrons: if n = 2, electrons dance solo in two styles. The fractional Hall effect in Fig. 6.8b shows the "dancing together" mode of electrons: if n = 1/2, two electrons dance together in one style; if n = 1/3, three electrons use one style to dance together; if n = 2/5, it is more complicated, 5 electrons dance together in 2 styles.

Take a closer look at the candied haws in Fig. 6.8, we realize that this model is related to topology.

The difference between fractional quantum Hall Effects (n = 1, 1/2, 1/3…) can be intuitively described by the difference in collective motion patterns of these ground states. Each fractional Quantum Hall State corresponds to a collective dance mode. Several simple patterns can be characterized by the number of "genus" in the topology introduced at the beginning of this chapter, as shown in Fig. 6.9.

Through the interesting model of candied haws, we have initially seen the correlation between electronic dance and topological images in graphene. However, "revolving dance" refers to the dance of electrons in space. The most exciting electronic dance in 2-d materials is jumped out through "spin". More precisely, it is the mode of "rotation plus revolution", or described in physics terms, called "the coupling of spin and orbit." This dance has a closer and more essential relationship with topology.

6.6 The Spin Dance of Electrons

When we introduced quantum mechanics in Chap. 2 of this book, we described the basic properties of electron spin.

Electrons have three important properties: mass, charge, and spin. The mass of electrons moving in crystals is replaced by effective mass. For graphene, the effective mass near the Dirac point is zero, resulting in ultrahigh electron transport properties in graphene, which has been introduced in the previous chapter. The charge of electrons is an important physical quantity. Electrons (or holes), as charge carriers, move under the action of an external electric field to form an electric current, thus becoming the basis of the work of electronic components. As for the spin of electrons, there are fewer engineering applications.

Although the development and application of electronics has a history of more than one hundred years, what is used and studied in circuits and electronic devices is basically only current, the flow of charge, which has nothing to do with spin. For decades, electronic technology has improved tremendously, but the pursuit of better and smaller devices never end. When Moore's Law was first introduced, it was an inspiring and refreshing prediction. More than 40 years later, it seemed to become a curse on electronics. Moore's Law predicts that the silicon-based semiconductor industry was almost at end of its life! In recent years, through the discovery of the giant magnetoresistance phenomenon, people have learned about the spin of electrons. The enthusiasm of scientists and engineers studying electronic technology are renewed. The electrical charge has dominated the industry for over 100 years, now, it is time for electron spin to take the center stage. Researchers hope to use the newly found mysterious nature of electronics to overcome bottlenecks, remove obstacles, and usher in a new era in the semiconductor industry. As a result, this has led to a large number of theoretical creation and experimental research on spintronics in recent years. This emerging discipline attempts to develop the spin transport properties of electrons. The spin is controlled by the magnetic field and electric field to generate a spin-polarized current, thereby increasing the freedom of electron movement in the external field, which can carry more information than a single charge, and enable the possibilities to manufacture smaller and faster electronic components than existing electronic components. Spintronics is of great value for both application and theoretical research. Quantum Spin Hall Effect is unique for graphene and topological insulators, which we will introduce soon. These materials have the potential to make significant progress in research of spintronics.

Let's revisit the basic concepts of electron spin. Electrons are particles with spin 1/2. This shows that electrons dance in two styles, just like a ballerina spinning around herself, clockwise, or counterclockwise. We generally use "up" and "down" to represent these two styles, as shown in Fig. 6.10a. Spin of an electron can be equivalent to a small magnetic dipole. Two spins "up" and "down" corresponding to magnetic dipoles with opposite polarities, which analog to ballerina turning clockwise and counterclockwise.

How electron spins interact with crystal lattice? This can be seen from Fig. 6.10b, c. These two diagrams are energy band density diagrams of spin state in non-ferromagnetic metals and ferromagnetic metals. Band density diagram is different from band structure but related. In Fig. 6.10b, c, the left half is energy band of "down" spin electron and the right half is energy band of "up" spin electron.

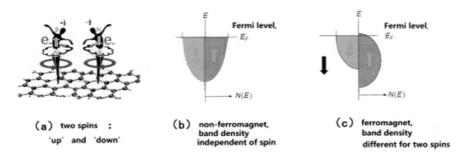

(a) two spins :
    'up' and 'down'

(b) non-ferromagnet,
    band density
    independent of spin

(c) ferromagnet,
    band density
    different for two spins

Fig. 6.10   The nature of electron spin

For non-ferromagnetic metals (Fig. 6.10b), the band density has nothing to do with spin orientations. That is because spin does not directly interact with crystal lattice if there is no magnetic field. When current pass a non-ferromagnetic metal, there is only electric field without any magnetic field. Therefore, there will be no difference in the resistance experienced by the electrons in different spin states due to lattice scattering. It is like a ballerina turning clockwise and another ballerina turning counterclockwise, dancing quickly between many fixed soldiers, bumped by the soldiers indiscriminately, and treated equally, no one cares if she is turn clockwise or counterclockwise.

However, it is different in ferromagnetic metal (Fig. 6.10c), where the soldiers themselves are also spinning rapidly, clockwise or counterclockwise. When soldiers and ballerinas who are turning in the same direction meets everything is in unison, and when the soldiers meet a dancer who turns in the opposite direction, chaos erupts. Spin dipoles in electronics are similar to ballet dancers. They interact with the magnetic moment (soldier) in ferrimagnets, which makes the density of states of two spin electron energy bands different. It can be seen from Fig. 6.10c that the energy bands of the two spin states in ferromagnetic material have shifted, showing left–right asymmetry. Especially near the Fermi level, the number of electrons whose spin orientation is consistent with the magnetization direction is relatively large, while the number of electrons whose spin orientation is opposite to the magnetization direction is very small, almost zero.

Because two different spin orientations have different performances in magnetic materials, this can be used to separate the spin electrons of different orientations to obtain a "spin current" with a single spin orientation.

It can also be compared with the original "charge current": the action of the electric field causes the positive and negative charges to move in different directions to form the so-called current; under certain conditions, the combined effect of the electromagnetic field can make the "up and down" spin electrons follow in different directions to form a spin-polarized current. The superiority of this spin current can be a way to revitalize the electronics industry, which is the original intention of spintronics research.

In a nutshell, the charge current is closely related to the electric field, while the spin current and the magnetic field are more closely related. For example, in our existing computer technology, the logic operation part uses electric current and the storage part uses magnetism, but the two are basically separated. If the charge current and spin current are combined and applied together, the function of the human brain can be better simulated, and the computer made in this way should be faster and more effective.

Spintronics uses charge plus spin to obtain a 4-state system, which has a higher data transmission speed than a 2-state system, improves processing capacity and storage density, and increases storage capacity. Spin provides an extra degree of freedom for storing and handling information, it may also be used in quantum computing and quantum communication devices to realize a completely different computing technology.

The single-atom layer structure of graphene has simple and adjustable spin characteristics and may be a promising material in spintronics. A team in the Netherlands has been studying the spin transport of graphene at room temperature since 2007. Rewarding progress has been made in graphene magnetization tunable and spin lifetime extension. It is planned to be applied to integrated circuits to open the door to graphene spintronics.

6.7   The Hall Effect Family

The difficulty in practical application of the quantum Hall effect is that it requires a very strong magnetic field. Among the classic family members of the Hall Effect, there are also two members that do not require an external magnetic field. One is the abnormal Hall Effect that Hall himself observed in ferromagnetic materials three years after the discovery of normal Hall Effect. The second is the spin Hall Effect that was predicted by theory very early, but was not experimentally confirmed until 2004. Now that there are two classical members that do not require a magnetic field, it should also be possible to observe their quantum counterparts: quantum anomalous Hall Effect (QAHE) and quantum spin Hall Effect (QSHE).

Quantum Hall effect without a magnetic field was predicted by American physicist Duncan Haldane in 1988. Later, Professor Charles Kane of the United States first conceived QSHE theoretically in 2005. Kane is also believed QSHE to be possible in single-layer graphene samples. Professor Shoucheng Zhang proposed in 2006 that it is possible to realize QSHE in the HgTe/CdTe quantum well system.

Later studies have shown that the spin–orbit coupling in graphene is very small, and it is difficult to observe QSHE. QSHE in the HgTe/CdTe quantum well system predicted by Zhang Shousheng was quickly confirmed by experiments by a German research team. Later, Spin Hall Effect was also observed in graphene and silylene. The team led by academician Xue Qikun of the Chinese Academy of Sciences discovered QAHE for the first time in the world in 2013.

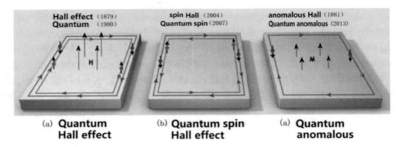

Fig. 6.11   Trio of the Hall effect family

The advantage of QSHE is that it does not require an external magnetic field, as shown in Fig. 6.11b. The sufficiently large spin orbits are coupled with each other to replace the effect of external magnetic field and generate fringe currents. The edge current here is different from the edge (charge) current of the quantum Hall effect described in Sect. 6.5, which is the spin current. In QSHE, electrons have 2-spins: up and down, which produce two opposite motions, the total charge current is 0, but the net spin current at the edge is not 0. The electrons "dance" very orderly in a new posture. The up-spin electrons and the downspin electrons move face to face, but each has its own way and does not interfere with each other, resulting in two spin currents.

What is the relationship between quantum spin Hall state and topology? To understand this, it is best to introduce topological insulators first.

6.8   Graphene and Topological Insulators

Quantum Hall effect and quantum spin Hall Effect introduced above both form edge currents. Another type of state closely related to the edge current is the topological insulator.

In a sense, it can be said that topological insulators are developed on the basis of QSHE, its concept can be extended to 3-dimensional materials. The most intuitive property of a topological insulator is that its interior is an insulator, but the surface can conduct electricity. It is like an insulated porcelain bowl. After being plated with gold, it has conductivity on the surface. However, they are two fundamentally different surface conductivities. The surface conductivity of gold-plated bowl is external to the porcelain and will disappear as the coating is damaged. The surface conductivity of a topological insulator is not the nature of surface, but the intrinsic nature of insulator body, so impurities and defects will not affect it.

In other words, the root of the conductive properties of the insulating surface in a topological insulator is from band topology of bulk material, not because the surface is coated with a certain conductive material. Cut off the original surface, the new surface will still conduct electricity, because the band structure of material will not change, and its topological properties protect the surface conductivity forever.

What exactly is band structure of topological insulators? Since it is an insulator, shouldn't the band structure be like the one drawn in Fig. 4.12 with a wide gap between the upper conduction band and the lower valence band?

Topological insulators are similar to ordinary insulators, and the energy gap between the conduction band and the valence band is very wide. What distinguished one from the other is the topology of energy bands is different. For example, band topology of an ordinary insulator is a ring as shown in the lower right figure in Fig. 6.12, while the conduction band and valence band of a topological insulator are entangled with each other, as shown in the lower left figure in Fig. 6.12. The knot cannot be opened as shown in the Figure. The specific shape of the knot and the reason for its formation may be different due to different materials, but the knot and the loop have a completely different topology. It is impossible to transition to the shape of an ordinary loop without cutting the knot and reconnecting it.

An example of so-called topological differences is band inversion. For ordinary crystal materials, s-orbitals split to form a conduction band, p-orbitals form a valence band, conduction band is above and valence band below. Under certain circumstances (such as HgTe predicted by Zhang Shousheng), the strong spin–orbit coupling effect pushes some of p-orbitals above s-orbitals, thus forming a band inversion.

Figure 6.12 visually shows the formation of the surface current of the topological insulator with the band inversion. The shaded part in the figure represents a topological insulator, and the un-shaded region is an external vacuum or a common insulator (vacuum is a common insulator). The boundary between shaded and un-shaded represents the surface of the topological insulator. The valence band is represented by a

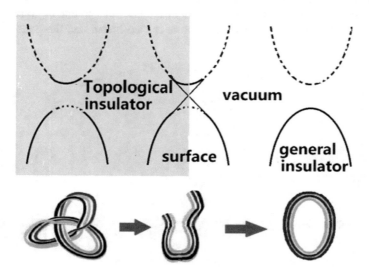

Fig. 6.12 Topological insulator

solid line, and the conduction band is represented by a dashed line. In the ordinary insulator in the upper right figure, the solid and dashed lines are completely separated, while in the top left figure, the top of valence band and bottom of conduction band, a section of solid line and dashed line are interchanged, which indicates the band inversion inside the topological insulator. The left side of interface is energy band of topological insulator with inversion, right side is ordinary normal energy band. How does the band diagram change to transition from inverted band to normal band? Just like a knot becomes a loop, it must be cut at the interface and reconnected. For the energy band diagram, two diagonal lines are added between conduction band and valence band, which means electrons on the interface have a transition from valence band to conduction band. This is the reason why the surface of topological insulator is conductive.

The topology mentioned by topological insulators has nothing to do with the topological shape of the material itself in real space, and has nothing to do with the spatial configuration of the crystal. It is the topology of band structure in wave vector space. The band diagram of interface in Fig. 6.12 appears familiar. Isn't that the Dirac cone in the graphene energy band diagram? In fact, it is precisely because of the special band structure of the graphene (Dirac cones), inspires the thinking of physicists, making them looking for quantum spin Hall states in graphene.

Figure 6.13 shows band diagrams of graphene (a) and several other states near the Fermi level.

Figure 6.13b is band diagram of quantum Hall state (or topological insulator). The shape of its conduction band and valence band near the Fermi level is close to a parabola, similar to ordinary insulators, but conducts electricity due to the existence of edge states. In Fig. 6.13b, the edge state of the Quantum Hall State is a straight line connecting the conduction band and the valence band. Therefore, the behavior of the Quantum Hall State near the low-energy state is similar to that of graphene. The relationship between energy and momentum is also linear, and there are massless

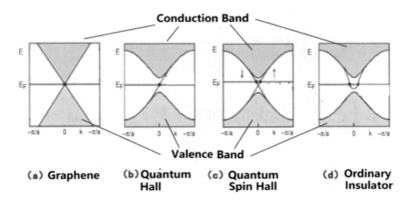

Fig. 6.13 Comparison of band structure for graphene and others

relativistic quasiparticles. Figure 6.13c is band diagram of QSHE, it is similar to Fig. 6.13b but there are 2 lines connecting conduction band and the valence band.

Ordinary insulators may also produce edge states to form edge conduction, but they are essentially different from the edge states of topological insulators. Figure 6.13d shows the band of an ordinary insulator. The edge state curve in the figure intersects Fermi level, which means that there can be an edge current in the insulator. This edge conductivity is unstable, edge state curve can be retracted to disappear because there is no topological protection. In two quantum effects shown in Figure b and Figure c, straight lines from top to bottom tight conduction band and valence band together. It can also be summarized in one sentence: ordinary insulators and topological insulators have different topological structures in edge states. The former is trivial topology, while the latter is no-trivial topology. The conductivity of topological insulators is protected by its topological properties.

Chapter 7
New Materials

7.1 Graphene Production Technique

Graphene was first discovered by peeling off a layer of graphite with tape. The method is simple but not a precise way to get large volume of high-quality graphene.

Today there are many methods to produce graphene, in general, they can be divided into physical methods and chemical methods. The two categories can also be divided according to the process: top-down and bottom-up graphene synthesis. The former refers to the separation or decomposition of graphite to obtain graphene; the latter refers to the covalent connection of small molecules based on chemical reactions to construct a 2D network structure, or graphene is built up from small to large. In Geim's days, he used a "top-down" physical stripping method.

7.1.1 Micromechanical Peeling Method

Sticking tape and peel can fall into the micromechanical peeling category. Mechanical peeling techniques also include ultrasound, ball milling, and using fluid dynamics.

When the Geim team claimed to stick graphene with tape, it was not a simple procedure. In fact, they first used an oxygen plasma beam to etch a groove surface with a width of (20 μm - 2 mm) and a depth of 5 μm on the surface of the highly oriented pyrolytic graphite (HOPG), which was pressed on a surface of SiO_2/Si substrate with photoresist. Then, after firing, the scotch tape is used to repeatedly peel off the excess graphite flakes, and finally the graphene layer with a thickness of only a few layers is obtained.

The micromechanical peeling method is relatively simple and able to produce high quality graphene downsides are low efficiency, high cost, and ability to get finer layers of graphene. The micromechanical peeling method is deemed not suitable for mass production.

© Guangxi Science & Technology Publishing House 2022
T. Zhang, *Graphene*, https://doi.org/10.1007/978-981-16-4589-1_7

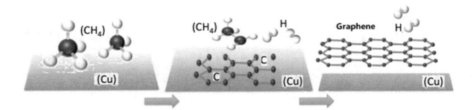

Fig. 7.1 Chemical vapor deposition method

7.1.2 Chemical Vapor Deposition (CVD)

Chemical vapor deposition is the most widely used method to produce large-scale industrial semiconductor thin film materials. In typical CVD, the wafer (substrate) is exposed to a gaseous condition, and undergoes a chemical reaction, then decompose on the substrate surface to produce the desired deposit.

For example, let the heated methane gas (CH_4) pass through the surface of copper foil (Cu). The copper foil separates carbon and hydrogen atoms from the methane. The carbon atoms are deposited on the surface of the copper foil to form a two-dimensional graphene with a hexagonal structure lattice. Finally, remove copper foil to obtain a single-layer graphene. This method has the potential of preparing graphene sheets with a larger area, as shown in Fig. 7.1.

South Korea's Samsung Electronics has used this method to obtain a single-layer graphene with a diagonal length of 30 in., demonstrating the feasibility of mass production method. In 2017, the Argonne National Laboratory of the U.S. Department of Energy developed a technology that uses diamond substrate materials to prepare single-crystal graphene by ultra-high-speed chemical vapor deposition (CVD). Japanese researchers have developed CVD method making graphene directly grown on plastic or silicon dioxide substrates at low temperatures. These developments have improved the quality of graphene prepared by chemical vapor deposition and greatly reduced the cost of graphene films.

Graphene grown on copper foil cannot be used directly. It must be peeled off and transferred to the required substrate, usually a plastic film. It is conceivable that it is not easy to tear off graphene with only one atomic layer and put it on other substrates. This is also a huge challenge for the practical use of graphene films. It is said that the Spanish Graphenano company is the world's most successful company in graphene transfer [21].

7.1.3 Epitaxial Growth Method

Epitaxy method is a method of growing a layer of another crystal with certain requirements on a lattice structure through lattice matching. This method is like extending

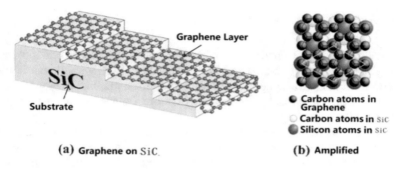

(a) Graphene on SiC

(b) Amplified

Fig. 7.2 Schematic diagram of epitaxial growth method

the original crystal by one layer, so it is called epitaxial growth. Compared with other methods, the graphene obtained by the epitaxial method has better uniformity, and because the epitaxial growth technology was originally developed in the production process of integrated circuits in the 1950s and 1960s, it was prepared by this method Graphene is compatible with IC technology.

The epitaxial method can be classified according to the different substrates selected. Figure 7.2 is a schematic diagram of the epitaxial growth method to grow a graphene layer on SiC.

In the 1990s, it was discovered that graphitization of SiC single crystals occurs when heated to a certain temperature. In 2004, the research team of Professor Walter de Heer from Georgia Institute of Technology in the United States proposed for the first time to synthesize graphene by SiC epitaxy. They used SiC single crystal as the substrate and used the etching effect of hydrogen on SiC at high temperatures to form a step array with atomic level flatness on the surface of the substrate. Then, in an environment with an ultra-high vacuum and a surface temperature of 1400 °C or higher, the C atoms on the surface of the substrate are reconstructed to form a hexagonal honeycomb-shaped graphene film, as shown in Fig. 7.2.

In addition to choosing silicon carbide (SiC) as the substrate, a metal-catalyzed epitaxial growth method is commonly used to prepare graphene. A transition metal substrate with catalytic activity (such as copper) is used. Hydrocarbons are introduced into the metal surface under ultra-high vacuum conditions. The carbon atoms on the copper surface are adsorbed and dehydrogenated, and arranged into a lattice to obtain graphene.

Compared with SiC epitaxy, metal epitaxial graphene is mostly single-layer, high quality, and easy to transfer. However, the morphology and properties of graphene prepared by this method are greatly affected by the metal substrate. Therefore, the choice of epitaxial substrates is not limited to these two types, and developers all over the world are still working hard on research and exploration.

7.1.4 Synthesis of Reduced Graphite Oxide Method

In this method, graphite is firstly oxidized with concentrated sulfuric acid, potassium permanganate, hydrochloric acid and other oxidants to form graphite oxide (GO). Then GO reduced to graphene flakes. The method from graphite to graphite oxide has long existed, dating back to the work of Oxford University chemist Benjamin Brody in 1859.

The specific operation process is shown in Fig. 7.3. First, in the presence of a strong oxidant, the graphite powder is oxidized and mixed, and then magnetically stirred to prepare graphite oxide. The difference between graphite oxide and graphite is that after graphite is strongly oxidized, it contains carboxyl groups, hydroxyl groups, etc. at the edges of its layers, while the interlayers contain oxygen-containing groups such as epoxy and carbonyl groups. The insertion of these groups increases the distance between graphite layers to more than twice the original distance, from 0.34 nm to about 0.78 nm. The second step is to dissolve the graphite oxide after mechanical separation (such as ultrasonic) treatment in water or other organic solvents, so that the graphite oxide can easily be dispersed in the solvent into a uniform single-layer or double-layer graphene oxide solution. Although graphene oxide has been exfoliated into a single-layer sheet, it still contains a lot of remaining oxygen-containing groups. In the last step, it is treated with a strongly alkaline and hygroscopic deoxygenated liquid such as hydrazine hydrate, and finally a fully peeled, single-layer or few-layer graphene is prepared.

The graphite oxide reduction method is one of the most popular methods for graphene. From large block structure of graphite, to graphite powder, to graphite oxide, and finally to single-layer graphene, the method is simple and low-cost. Large quantities of graphene can be produced. Its disadvantage is the quality of final product. Graphene that has been completely oxidized by a strong oxidant may not be completely reduced to a single-layer graphene with a perfect crystal structure, which will cause some of its physical and chemical properties to decrease.

Although it is difficult to obtain pure graphene as the final product of this method, the different by-products and derivatives obtained in the whole process also rouse the interest of researchers. For example, graphene oxide GO with a large number of

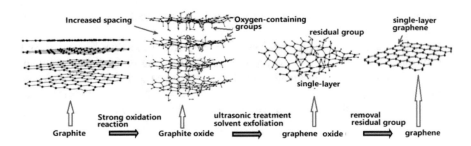

Fig. 7.3 Schematic diagram of graphite oxide reduction method

oxygen-containing functional groups on the surface has many special properties and has developed many practical applications. In addition, the graphite oxide reduction method also provides a theoretical basis and experimental flexibility to synthesis other nanocomposite materials that are not exactly the same as graphene.

7.1.5 Electrochemical Method

The electrochemical method of graphene is another "top-down" chemical method. Starting with massive graphite, using chemical methods to increase the distance between graphite layers, finally crack and disperse into Graphene sheets.

Insert two (or one) high purity graphite rods into the aqueous solution containing ionic electrolyte, as shown in Fig. 7.4. The cations and anions in the ionic liquid move to cathode and anode respectively under electric field force, and are inserted between graphite layers, so that the graphite layer spacing is increased. The interlayer force is reduced so that ions and π electrons in the layer are combined. Therefore, graphite rod is gradually corroded to form a functionalized graphene sheet. Finally, it is washed and dried with absolute ethanol to obtain graphene. However, the graphene sheet prepared by this method is more than a single atomic layer.

7.1.6 Ultrasonic Liquid-Phase Exfoliation Method

Similar to the electrochemical method described above, it also mainly uses graphite blocks as the initial material. First, it is dispersed in a solvent. When the solvent molecules are adsorbed on the graphite surface, ultrasonic vibration is used to pull the graphite in the liquid phase. The surface layer is peeled off to form graphene.

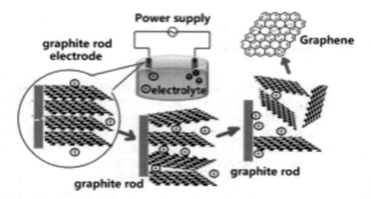

Fig. 7.4 Schematic diagram of electrochemical method

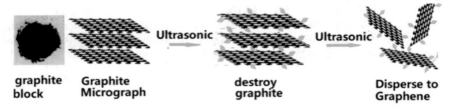

| graphite block | Graphite Micrograph | | destroy graphite | Disperse to Graphene |

Fig. 7.5 Schematic diagram of ultrasonic liquid-phase exfoliation

Immediately after the new graphite surface adsorbs solvent molecules and continues to be peeled off, the graphite block is continuously peeled from the outside to the inside during the whole process, just like peeling an onion, graphene sheets are continuously produced. Because it uses the mechanical action of ultrasonic waves, there are fewer chemical oxides formed, and the quality of graphene is higher than that of electrochemical methods. However, it takes a long time to peel off using only ultrasonic vibration, and long-term force will vibrate the graphene into small pieces and multiple layers. Therefore, this method can be mixed with other preparation methods (Fig. 7.5).

7.2 Graphene Family Nanomaterials

The term graphene originally refers to a two-dimensional crystal with a single-layer carbon atom structure. This hexagonal benzene ring structure composed of carbon atoms has many other variations, each with their own unique properties and In this section, we are going to review variants of graphene that still maintain the hexagonal structure, but are not ideal single-layer graphene, which can be regarded as close relatives of two-dimensional graphene.

7.2.1 Multilayer Graphene

Single-layer graphene has many unique l electrical, optical and mechanical properties. Strictly speaking, only single layer of atoms is true graphene. In general, a crystal structure with less than 10 layers of carbon atoms is generally referred to as graphene. However, the band structure and properties of multilayer graphene are different from those of single-layer graphene. In the study of various layers of graphene, many fascinating properties were found. Especially under controllable conditions, multilayer graphene composed of a certain stacking method (such as AA, AB, ABC, or ABA stacking, etc.) has both flexible application value and interesting theoretical significance. See Figs. 7.6 and 7.7.

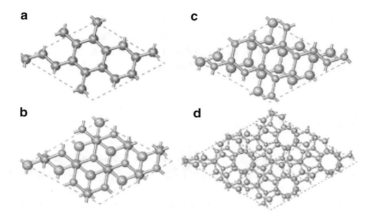

Fig. 7.6 Schematic diagram of double-layer graphene

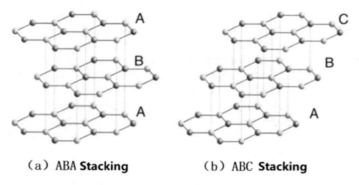

Fig. 7.7 Two different stacking methods of three-layer graphene

There can be many different overlapping modes between the two layers of double-layer graphene. For example, the lattices of the two layers are completely overlapped, which is called AA mode, as shown in Fig. 7.6a; or lattice atoms of the upper graphene sheet are located just at the center of the hexagon of the underlying lattice (AB mode), as shown in Fig. 7.6b. Between the two graphene sheets shown in Fig. 7.6c, the crystal lattice has a parallel displacement; the crystal lattice of the two sheets of graphene shown in Fig. 7.6d rotates at an angle with respect to each other. The double-layer graphene stacked by different modes have different properties. For example, AB mode graphene has a band structure that is not a Dirac cone, but returns to a bowl-shaped parabola. The carrier is a chiral fermion with mass, and its properties are very different from single-layer graphene.

For three-layer graphene, its band structure is more complicated and its properties are also different. Figure 7.7 shows two stacking methods of 3-layer graphene. Whether it is double-layer AA, AB, or three-layer ABC, ABA and other stacking methods, it can be repeated many times to form a more layered structure. In fact, graphite, which readers are familiar with, is made up of repeated stacking of multi-layer graphene (ABAB…). But in the material that makes up pencil lead, there are more than 10 layers, so it can only be called "graphite", not "graphene".

7.2.2 Graphene Nanoribbons

Graphene film has many special properties, which raises an interesting question: What about the properties after cutting it into strips? When the width of such a graphene strip is less than 50 nm, it is called a graphene nanoribbon.

Because the width of graphene nanoribbons is at the nanometer level, it can be regarded as a one-dimensional material. The theoretical model of nanoribbons was first proposed in 1996. After graphene has become a star, nanoribbons have naturally become a research focus. Due to its variable width and rich edge configuration, it has many properties and applications that are different from two-dimensional graphene.

Cut the graphene into a ribbon shape, and the cut edge shape can be divided into a zigzag shape and an armchair shape, as shown in Fig. 7.8.

The tight-binding approximate model can be used to calculate the band structure of the nanoribbons. The zigzag shape is metallic, and the armchair shape is metallic or semiconducting, depending on the width of nanoribbons. Research results show that the semiconducting band gap is inversely proportional to width of nanoribbon. In other words, when the armchair-shaped nanoribbon is wider, the energy gap is smaller and the metallicity is stronger. This is not difficult to understand, because the wider nanoribbon is closer to two-dimensional graphene.

Nanoribbons inherit many excellent properties of graphene, such as high electrical conductivity, high thermal conductivity, and low noise. These excellent qualities have prompted graphene nanoribbons to become another choice for integrated circuit interconnection materials, possibly replacing copper metal. It is also possible to make electronic devices such as field effect transistors, lasers and amplifiers.

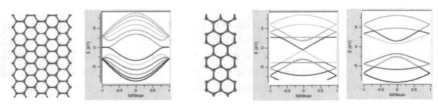

(a) **Zigzag nanoribbons**   (b)   **bands of Armchair nanoribbons**
    bands metallicity   **from metallicity to semiconducting**

Fig. 7.8 Graphene nanoribbons (from Wikipedia)

7.2.3   Carbon Nanotubes

Another one-dimensional nanomaterial among the close relatives of graphene is carbon nanotubes that predate graphene. Carbon nanotubes were discovered in carbon fibers produced by the arc method using a high-resolution transmission electron microscope by physicist Sumio Iijima in 1991. It is actually equivalent to graphene nanoribbons just introduced by curling up in the width direction. Nanotubes are only nanometers in radial direction, but can be as long as tens to hundreds of microns in the axial direction, so they are called carbon nanotubes.

If, as just said, nanotubes are formed from nanoribbons crimped, because nanoribbons are divided into zigzag and armchair shapes, there are two types of nanotubes. After the nanoribbons with the zigzag edge are curled, its port is in the shape of an armchair, which is called an armchair carbon nanotube, as shown in Fig. 7.9c; while the nanoribbons with the armchair-shaped edge are called zigzag carbon after curl Nanotubes, see Fig. 7.9b. Because the names of the two types of carbon nanotubes and nanoribbons are just reversed, their conductive properties are reversed from the description in Fig. 7.8: armchair-shaped carbon nanotubes always appear metallic.

There are more than two types of carbon nanotubes. From the perspective of cutting two-dimensional graphene into strips and then curling up, there can be more types of curling directions. As shown in Fig. 7.9a, in addition to curling in the horizontal direction to obtain armchair-shaped carbon nanotubes, and curling at a 45° angle to obtain zigzag carbon nanotubes, it can also be curled in any other angle direction. The properties of the obtained carbon nanotubes may be different from the above two, and they are called spiral type. Generally speaking, carbon nanotubes composed of different crimping angles, corresponding to different combination values (n, m), can be used as an indicator of the shape of carbon nanotubes.

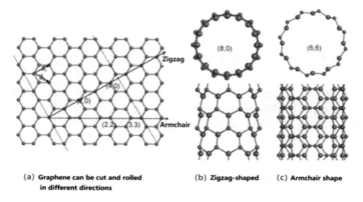

(a) Graphene can be cut and rolled   (b) Zigzag-shaped   (c) Armchair shape
     in different directions

Fig. 7.9   Carbon nanotubes

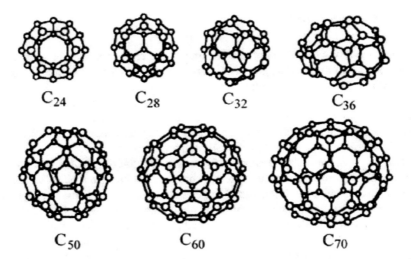

Fig. 7.10 Example of fullerene (buck ball) (https://www.sciencedirect.com/topics/medicine-and-dentistry/fullerene-derivative)

7.2.4   Fullerene

The hexagonal lattice of graphene can also form a 0-dimensional structure, such as the fullerene mentioned earlier. It is a hollow molecule composed entirely of carbon, and its shape can be spherical, ellipsoidal, cylindrical, tubular or other. Fullerene is very similar to graphite in structure. Graphite is made up of graphene layers composed of six-membered rings. Fullerene not only contains six-membered rings, but can also have five-membered or seven-membered rings (Fig. 7.10).

7.3   Derivatives of Graphene

Derivatives produced during the preparation of graphene, or incompletely pure graphene produced by deliberate doping or other methods, sometimes exhibit some unexpected properties. Here are a few examples.

7.3.1   Graphene Oxide

The graphene oxide in the intermediate process of producing graphene by oxidation and reduction described above is widely used because of its special permeability. In other words, graphite is strongly oxidized with concentrated sulfuric acid, potassium permanganate, hydrochloric acid and other oxidants to form graphite oxide, and

Fig. 7.11 The spin arrangement of graphene makes it show magnetism

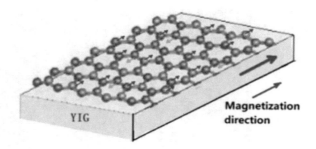

then uniformly dispersed in water by ultrasonic treatment, and finally thin slices of graphene oxide (GO) with controllable area and thickness can be realized. The method is simple to produce GO film, which has the advantages of large area, high mechanical strength, uniformity, etc., and is easy to assemble in an infiltration device. It is a promising environmental purification material.

7.3.2 Magnetic Graphene

Graphene has chemical stability, high conductivity and strong strength. In general, they are not magnetic. Electricity and magnetism are both indispensable for the development of ultra-efficient microprocessors in semiconductor technology, and magnetism is the foundation of storage elements in computers.

In 2015, scientists at the University of California, Riverside successfully developed graphene with magnetism. They placed an ordinary non-magnetic graphene film on the base layer of magnetic yttrium iron garnet (YIG), as shown in Fig. 7.11. As a result, they successfully transferred the magnetism of YIG to graphene without destroying its structure and other properties. Most conductive magnetic substances may interfere with graphene's super conductive ability, but because YIG is an insulator, it does not affect the electron transport performance of graphene. After experimental identification, the final magnetism of the graphene film is derived from the graphene itself, which is the result of the electron spin orientation in the graphene being changed by the magnetic substance.

7.3.3 Graphene Sandwich

The "sandwich" structure is of a familiar occurrence in the semiconductor industry. Shockley, one of inventors of transistor, designed his junction transistor into a sandwich bread style. Because there are P-type and N-type doped semiconductor materials, there are two ways to form a sandwich: NPN or PNP, corresponding to the two main transistor forms. The giant magnetoresistance effect discovered in the late

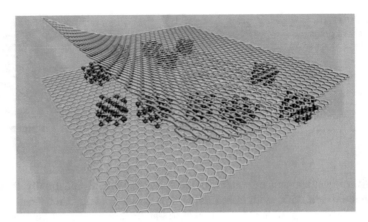

Fig. 7.12 Graphene sandwich (https://www.futurity.org/nanoscale-sandwich-graphene-137072 2-2/)

1980s also occurs in a sandwich structure composed of a magnetic metal and a non-magnetic metal.

Scientists naturally "play" the sandwich method to graphene. Figure 7.12 is an example.

This is a nano-scale sandwich structure designed for graphene by researchers at Rice University in the United States: two sheets of graphene are placed above and below the nanoclusters of magnesium oxide. They conducted computer simulations on this material, and the results showed that the magnesium oxide nanoclusters sandwiched between graphene layers formed compounds with unique electronic and optical properties, which may be suitable for sensitive molecular sensing, catalysis and biological imaging.

7.3.4 Calcium Doped Graphene

Adding some other atoms to graphene will change the properties of graphene. This is also one of the ways that materials scientists "play" with graphene. Researchers from Tohoku University and Tokyo University in Japan use two sheets of graphene to build a sandwich-like structure. After inserting some calcium atoms in the graphene sheet, it was surprisingly discovered that this structure achieved superconductivity! In other words, the resistance of the material so constructed is zero (Fig. 7.13).

When electrons pass through graphene without resistance, it is finally possible to manufacture energy-saving and high-speed nanoelectronic devices. Unfortunately, the superconductivity of calcium-doped graphene occurs at around −269°C (4 K). At ultra-low temperatures, the conductivity of this material drops rapidly and superconductivity appears.

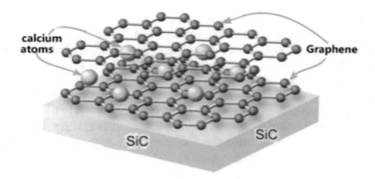

Fig. 7.13 Graphene shows superconductivity after adding calcium atoms (https://phys.org/news/2016-02-graphene-superconductiveelectrons-mass-resistance.html)

Research on graphene has shown that it is possible to obtain the dawn of "high temperature superconductivity." What is high temperature superconductivity? At what temperature is the high-temperature superconductivity occurring? In the next section, we will first review the history of superconductivity and then describe an interesting research result of MIT scholars in the double-layer graphene experiment in 2017: the discovery of a different general superconductivity phenomenon in graphene.

7.4 Superconductivity and Graphene

As we all know, materials consume energy in the process of conducting electricity, which is manifested as resistance of the material. The greater the resistance, the more energy is consumed. Generally speaking, the resistance decreases as the ambient temperature decreases. In 1911, Dutch physicist Heike Onnes (1853–1926) discovered that mercury's resistance equals to 0 at low temperatures (about 4 K), which is called superconductivity phenomenon. Onnes won Nobel Prize in Physics in 1913 for this. Later, American physicists John Barding, Leon Cooper, and John Schriever proposed BCS theory named after the first letter of their name, explaining the microscopic mechanism of superconductivity, and this theory Known as the conventional explanation of superconductivity. The BCS theory believes that the vibration of the lattice makes two electrons with opposite spin and momentum form a Cooper pair with zero momentum and zero total spin. The Cooper pair, like a superfluid, can be bypassed. The lattice defect impurities flow to form a superconducting current without hindrance. The academic community believes that the BSC theory basically explains superconductivity at low temperatures. The three scholars won the Nobel Prize in Physics in 1972.

In short, superconducting materials have a critical temperature below which the resistance of the material is zero. However, the conventional superconductivity

explained by BSC theory generally occurs in a low temperature environment close
to absolute zero. Because BCS theory believes that Cooper pairs are produced by
condensation under low temperature conditions. Based on this explanation and experimental results, American physicist William L. McMillan predicted that there might
be an upper limit (about 40 K) for the transition temperature of superconductivity,
the so-called McMillan limit. The critical temperature of conductive materials may
all be below this upper limit.

The two basic characteristics of superconducting materials, zero resistance and
perfect diamagnetism, have made them suitable for many practical applications.
Materials with zero resistance can pass current without consuming energy, naturally they are the materials of choice to make electronic devices. Diamagnetism
is also known as the Meissner effect, which means that when a superconductor in
a superconducting state is placed in a magnetic field, the magnetization generated
inside it will completely cancel out the external magnetic field, thereby the internal
magnetic induction intensity Is zero. In other words, the magnetic field lines are
completely repelled from the superconductor, which is also the basic principle of
magnetic levitation in practical applications (Fig. 7.14).

Application areas of superconductivity include MRI in hospitals, accelerators,
magnetic levitation, and nuclear fusion research.

The low-temperature superconducting maglev train developed by Japan set a
world record of 603 km per hour for ground rail vehicles in 2015, and plans to build
the central Shinkansen maglev line in 2027, fully demonstrating the huge potential
of superconducting applications.

Superconductivity is one of the greatest scientific discoveries of the twentieth
century. The application of low-temperature superconductivity relies on expensive
cryogenic liquids, such as liquid helium, to maintain a low-temperature environment.

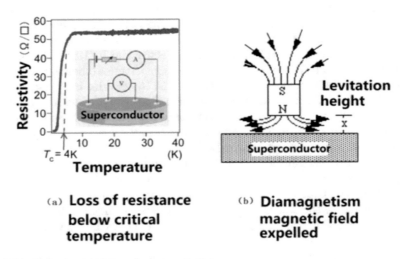

Fig. 7.14 Basic characteristics of superconductivity

This has led to a sharp increase in the cost of superconducting applications, making it more difficult to be widely used in large-scale engineering fields such as power transmission. The superconducting magnetic levitation trains described above are also expected to benefit from the emergence of high-temperature superconducting materials. Superconductivity has now been discovered for more than 100 years. For a long time, the so-called McMillan limit has become a major bottleneck restricting the wide application of superconductors even though it has not affected scientists' enthusiasm for superconductivity.

About 30 years ago, the roadblock of McMillan's limit was finally subdued, and experiments continued to discover way to exceed McMillan's limit.

The revolutionary breakthrough came from IBM in Switzerland. Swiss physicist K. Alex Müller and his students Georg Bednorz began working closely together in 1983 to conduct systematic research on superconducting oxides with high critical temperatures. In 1986, they discovered superconductivity with a critical temperature of 35 K in the ceramic material (BaLaCuO or LBCO), which was the highest record for the critical temperature at the time and broke the traditional belief "oxide ceramics are insulators", this have caused a buzz among scientists. Alex Müller and Georg Bednorz won the Nobel Prize in Physics in 1987. Material scientists flocked to use a variety of different compounds to search for better material and higher critical temperature.

In 1987, the research team of Chinese scientists Zhao Zhongxian and the teams of Zhu Jingwu and Wu Maokun of the United States independently worked on the Ba-Y-Cu-oxygen-based materials (Ba-Y-Cu). (-O) increased the superconducting critical temperature from 40 to 93.2 K, which greatly exceeded the McMillan limit. At the end of 1987, the record of critical superconducting temperature was increased to 125 K. Later, people referred to superconductivity at lower temperatures than the original liquid helium as high-temperature superconductivity.

The study of high-temperature superconductivity is still an important research topic in condensed matter physics. Three types of high-temperature superconductors have been discovered: copper oxide, iron-based, and magnesium diboride. However, conventional BCS theory cannot successfully explain the high-temperature superconductivity of these materials. Readers should also note that the so-called "high-temperature" superconductivity here is only compared to the low-temperature superconductivity of conventional superconductors, which is minus 270°C. The high temperature here can be as low as −196°C (77 K) liquid nitrogen. In fact, it is still the ultra-low temperature in our usual sense.

In 2019, scientists in Germany have hit a new superconductivity milestone—achieving a resistance-free electrical current at the highest temperature yet: just 250 K, or −23 °C which is currently the highest known Temperature superconductor.

High-temperature superconductivity and graphene are now two hot topics in the research areas of physics. It is inevitable that the two paths will cross and perhaps collide to generate exciting new discoveries. This is what happens to the 2017 MIT Condensed Matter Physics A wonderful discovery encountered by the research team of Pablo Jarillo-Herrero [22–24].

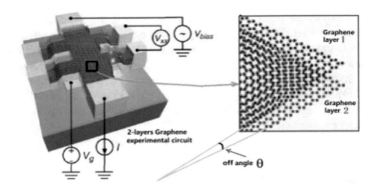

Fig. 7.15 Experiment of the relationship between the electrical properties of double-layer graphene and the relative deviation angle (from MIT article)

Initially, the MIT team was not to explore superconductivity. Their purpose was to explore how the deflection angle of double-layer graphene affects the performance of graphene, and designed an experiment for this purpose: stacking two layers of graphene sheets, however, the orientation of the lattice of two layers is rotated by an angle θ, as shown in Fig. 7.15.

When the angle θ is changed, electrical properties of the material will change. The researchers were surprised to find an unexpected behavior of double-layer graphene. When θ is exactly equal to a special angle $\theta = \theta_0(1.1)°$, the double-layer graphene material has superconducting properties. This result makes physicists excited.

Moreover, further studies have shown that the superconducting behavior of graphene is similar to that of copper oxide superconductors. Although the MIT team's superconductivity experiment results are still obtained at extremely low temperatures, they believe that this superconductivity of graphene could occur at room temperature because of its microscopic mechanism is consistent with superconducting materials, which is unexplainable using BCS theory. Therefore, whether it is high-temperature superconductivity requires further experimental verification.

Why is the superconducting behavior of graphene discovered by the MIT team similar to that of copper oxide? Although the microscopic mechanism of copper oxide superconductivity is still a mystery, and a complete theoretical framework has not been established, scientists have gained a lot of inspiration from the experimental results from studying them. Cooper pairs still exist in high-temperature superconductors, and the "critical" temperature of superconductors is actually determined by the density of electron pairs, that is, the degree of interaction between electron pairs.

The interaction between electrons in copper oxide materials is very strong, and its electronic behavior seems to be incomprehensible with the knowledge of quasiparticles and energy band theory based on Fermi liquids. Although high-temperature superconducting still appears due to the condensation of Cooper pairs, the main cause of Cooper pairs may not be due to electron–phonon coupling, but it may be dependent on the interaction between electrons, and related to Mott (Mott) Insulator.

The energy band theory of solids successfully describes the electronic properties of materials, allowing us to distinguish between conductors and insulators. However, there are always exceptions. The Mott insulator proposed by Nevill Mott and Rudolf Peierls in 1937 is an exception to the band theory. Mott insulator is a peculiar material. From the perspective of the energy band structure, its energy band is half-filled and should be able to conduct electricity. It is a conductor classified under the conventional energy band theory, but it is an insulator when measured at very low temperatures. The reason is due to the interaction between electrons. This is not considered in the conventional energy band theory.

The reason why Mott insulators exhibit insulation at low temperatures is due to the strong electrostatic interaction between electrons, which makes all electrons blocked and unable to flow. Under certain conditions, the system can become conductive and superconductivity appears.

Now look at the double-layer graphene used in the MIT experiment. Because the lattice of the double-layer graphene rotates at an angle, the lattice structure is changed. Figure 7.16a shows the Moiré pattern caused by the dislocation of the lattice. Figure 7.16b shows the change of the Brillouin zone in the wave vector space. The two small hexagons on the right in the figure are mini Brillouin zones.

Now let us look at the changes in the energy band diagram and briefly explain why superconductivity occurs at the magical angle of 1.1°.

According to the analysis of the MIT team, this magical angle can be calculated based on the change in the energy band diagram of the double-layer graphene with respect to the angle. When the graphene layers are twisted at an angle, the electron orbitals therein will re-hybridize and change the hybridization energy. The result of competition between hybrid energy w and electronic kinetic energy creates this magic angle. In other words, the torsion angle gradually increases, and the hybrid energy w increases. When Fermi velocity drops from $v_0 = 10^6$ m/s in the single-layer graphene to 0, the corresponding torsion angle is the magic angle. At this time, the

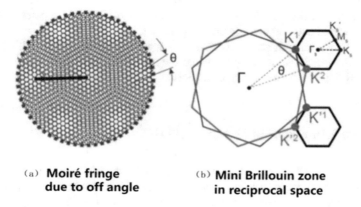

(a) **Moiré fringe due to off angle**   (b) **Mini Brillouin zone in reciprocal space**

Fig. 7.16 Double-layer graphene with inter-lattice rotation (from MIT article)

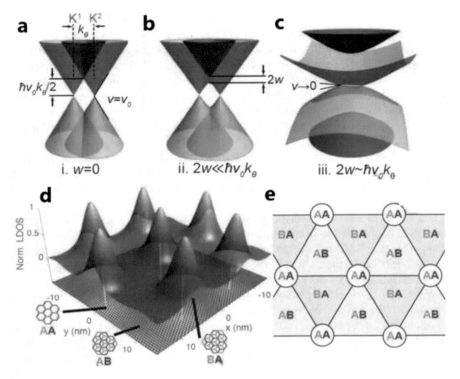

Fig. 7.17 Double-layer graphene superconductor (from MIT article)

corresponding hybrid energy is equal to the electronic kinetic energy, that is, $2w = \hbar v_0 k \theta_0$, and the magic angle $\theta = \sqrt{3}w\hbar v_0 K = 1.08°$, which is about 1.1°. The corresponding band diagram becomes an almost flat insulator band diagram, that is, a phenomenon similar to Mott insulators occurs. Insulation and superconductivity can be converted each other, only one-step away (Fig. 7.17).

If high-temperature superconductivity can be realized on a material with a simple structure like graphene, it has great value for application and theoretical research. This is why MIT's research excites physicists.

7.5　Two-Dimensional Nanomaterials

Today, materials, information and energy are known as the three pillars of modern civilization. The research and development of graphene has become hot topics in both academia and industry, pushing the production and manufacturing of various two-dimensional nanomaterials to a pinnacle.

Graphene is a layer of carbon atoms arranged in a 2-dimensional lattice, which produces various enchanting properties. Why other atoms cannot be arranged in a 2-dimensional lattice and have similar properties? Now days many cousins of graphene are emerging in the researcher's laboratory and the "family" is growing at an astounding speed! Silicene, germanene, stanene, boronene, nitrene, blue phosphorene, arsenene, antimonene, and many other 2-d materials have emerged.

What is the element that is most similar carbon? Scientists naturally think of silicon. It just happens that silicon is the most used basic material in our semiconductor industry. Silicon is the 14th element, and similar to carbon, outer layer also has 4 electrons, so that they can form hybrid orbital structures. As a result, materials scientists conceived of creating a 2-d hexagonal lattice of silicon atoms that is completely similar to the structure of graphene. This concept was formally proposed in 2008 and called silicene.

Scientists conducted research on simulating structure of silicon 2-d crystal materials for the first time in 1994 and researchers finally succeeded in manufacturing silicon with this structure in the laboratory in 2012, 18 years later. The process of producing silicene enabled and inspired advancement of graphene.

Start with the preparation methods, silicene and graphene share a lot of similarities, same with theory... Due to the original application characteristics of silicon materials, scientists have tried to apply silicene to electronic parts. Although graphene has a stronger conductivity than silicene, it lacks an energy gap. The energy gap needs to be opened artificially, such as applying an electric field, doping atoms, etc., before it can become a material for manufacturing logic transistors. Silicene has an advantage with the existence of a band gap, and its 2-d lattice is not as flat as graphene. Some atoms buckle up and some electrons are in slightly different energy states, forming a band gap. The weakness of silicene is that the preparation process is relatively complicated, and it is also difficult to separate from the substrate after film is made. In addition, silicon atoms are heavier and larger than carbon atoms, and large atoms affect the stability of chemical bonds. Therefore, silicene is very unstable when exposed to air. Researchers must find ways to overcome above difficulties.

The first production of transistors from silicene was achieved in February 2015 by a research team of Italy and United States. They first deposited silicon vapor on small silver crystals covered with aluminum oxide, and finally silicene was sandwiched between silver and aluminum to make a monoatomic silicene transistor. Silicene transistors are relatively small, and compared with current transistors, silicide transistors can improve chip performance and consume relatively less energy. Unfortunately this transistor was not stable at that time. When silicene is exposed to the air, it will begin to degrade in 2 min, and its life cycle is only a few minutes long. The technology of silicene transistor attracts the interest of scientists in both experiment and theory due to its potential performance and wide range of applications. There are also scientists who have broken through the difficulties to make silicene nanoribbons on silver substrates and then rebuild transistors. Even with those setbacks, silicene is still known as the most promising 2-d nanomaterial besides graphene.

Silicon and germanium are paired materials in semiconductor industry, like carbon and silicon, the number of electrons in the outermost layer of germanium is also

4. Therefore, in addition to silicene, many people also study germanene. The sp^3 hybridization of germanium is more stable than the sp^2 hybridization, and the length of the Ge–Ge bond is also longer than that of C–C, so it is more difficult to form germanene. However, germanene has been obtained through chemical deintercalation in the laboratory, and research has found Germanene is a kind of "2-d structure" material with very active chemical properties, it is also hopeful for Germanene to become the next exciting 2-d material. Theoretical analysis pointed out that 2-d germanene has greater spin–orbit coupling than graphene and silicone. It is possible to realize the quantum spin Hall Effect at room temperature, and it may become an vital basic material for information devices in the future. In 2014, researchers successfully prepared a single layer of undulating germanene using molecular beam epitaxy.

Carbon has many siblings in the periodic table, as shown in Fig. 7.18. The carbon group elements, as well as the boron group elements and nitrogen group elements adjacent to them, have become the research objects of scientists, and we will not be discussed in detail here.

Material science is bound to have a lot to do in future science and technology, not only 2-d materials, but also crystalline materials. This also reminds people of an application example from half a century ago, carbon fiber could only be used for fishing rods and golf rods, but it has now become an indispensable core material in aerospace field.

Nowadays, semiconductor industry is facing the threat of the "end" of Moore's Law, and it is critical to explore and find suitable new materials. Experts expect that the ultra-high electron mobility of graphene will make great applications in electronics industry in the future. The basic materials of integrated circuits now are semiconductors such as silicon or gallium arsenide. We learned from previous chapters that between conduction bands and valence bands of semiconductors, there is a small forbidden band (energy gap). Size of the energy gap determines characteristics of the semiconductor. Graphene is a zero band gap material, and the conduction band and valence band are connected by Dirac points. This unusual feature explains the

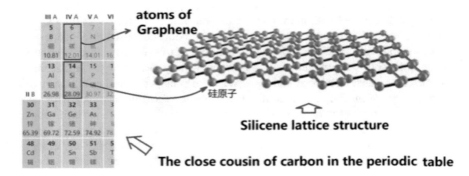

Fig. 7.18 Graphene's cousins

source of many magical properties of graphene, but it also greatly limits applications of graphene on electronic devices.

Question: can graphene be modified to form a certain and controllable energy gap in its band structure?

It sounds easy. Since the zero-energy gap is derived from the perfect crystal structure of graphene, can this perfect structure be disturbed to create an energy gap? In theory this is possible, but what method to use? How much energy is needed? Can the energy gap be created while maintaining the high-efficiency electron transport properties of graphene as much as possible? To answer those, many theoretical and experimental explorations have been developed around this purpose. It is generally believed that to make a graphene field effect transistor (FET) with a high switching ratio, opening the band gap is a necessary condition.

After several years of research, numerous methods for opening the energy gap have been developed, such as: using physical or chemical methods to destroy the crystal lattice, introducing doping or defects, using adsorbed atoms, adding external fields or stress, using the influence of the substrate, and using Spin–orbit coupling, etc..

Figure 7.19 shows the comparison of the energy band of double-layer graphene with that of single-layer graphene and the schematic diagram of the process of opening the energy gap of double-layer graphene with an external field.

Figure 7.19a shows the conical conduction band and valence band of single-layer graphene. They have no band gap and meet at one point. The energy band diagram of symmetrical double-layer graphene shown in Fig. 7.19b also lacks a band gap, but the energy band diagram is not cone-shaped and has a parabolic shape. In Fig. 7.19c, the direction indicated by the arrow is the magnetic field or electric field applied perpendicular to the graphene plane. The external field introduces asymmetry into the double-layer graphene, resulting in a controllable band gap (Δ), that is, the size of the band gap varies with the external field The strength is adjustable.

In fact, for single-layer graphene, taking into account the spin–orbit coupling effect, it is also possible to open the band gap with an external electric field, because the external electric field will bring about a momentum-dependent spin splitting effect, plus other mechanisms to strengthen the effect, it is possible to open the band

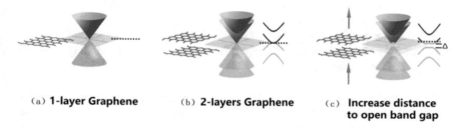

(a) **1-layer Graphene**　　(b) **2-layers Graphene**　　(c) **Increase distance to open band gap**

Fig. 7.19 Double-layer graphene opens the energy gap (https://newscenter.lbl.gov/2009/06/10/graphene-bandgap/)

gap. Experiments on the interaction of light and graphene show that light excitation also has the potential to open the band gap of graphene.

In summary, there are many theoretical predictions and experimental methods for opening the graphene band gap. It is up to the scientists and researchers to explore and narrow it down to one or a few practical methods that can be practical and applied to the existing technology.

7.6 Three-Dimensional Graphene

How to make more use of the excellent properties of graphene? Researchers have racked their brains. Some people have reduced 2-d materials to 1 or 0 dimensions, such as the carbon nanotubes and fullerenes introduced before. Others (MIT) have extended the 2-d to 3-d, claiming to make "3D graphene". The new 3-d material constructed has its own uniqueness after all. Therefore, we may wish to examine little bit: what is this 3D graphene?

Graphene film has special mechanical properties. As shown in Fig. 4.1a, graphene can be made into a very thin and light hammock because its strength is 100 times higher than the best steel in the world. However, steel is a three-dimensional material, and it is difficult to prepare large-area two-dimensional graphene. Even if it is available, it cannot be said to play a 3-d use like steel. So, is it possible to stack fragments of 2-d graphene or its derivatives into a 3-d material so that it inherits some of the powerful mechanical properties of graphene? The strength of graphene materials is derived from the strong σ bonds in the hexagon. As long as the 3D materials produced by stacking retain enough hexagonal σ bonds, they should be able to achieve certain strength.

Some people may say that if graphene is stacked into 3 dimensions, wouldn't it turn back to graphite? That may not necessarily be true; it depends on how you stack them. Graphite has absolutely no strength comparable to steel, because the graphene in graphite is arranged too neatly layer by layer. There is only a very unreliable π bond between layers. Such a regular pattern of overlapping playing cards makes the graphite soft and not rigid. The MIT scholars took a completely different stacking method. In the process of graphene synthesis, they increased heat and pressure, and used a 3D printer to compress small pieces of graphene together, resulting in a complex sponge-like structure similar to coral, a porous 3-d graphene material as shown in Fig. 7.20. According to a report by the MIT School in January 2017, the density of 3D graphene material is only 5% of steel, but the strength is ten times that of steel.

Interestingly, the MIT team realized a truth in the process of manufacturing porous 3D graphene. For the synthesis of new super-strong and lightweight materials, perhaps the most important thing is not the material itself, but the geometric model. They have developed a basic model. For which the materials does not even have to be limited to graphene, and other materials can be used.

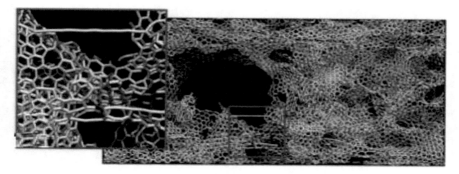

Fig. 7.20   3D graphene (https://newatlas.com/3d-graphene/47304/)

There are many researches on the application of 3D graphene, because the application range of monoatomic layer graphene film is limited after all. The University of California at Los Angeles in the United States and South Korean researchers have separately produced electrode materials using 3D graphene and tried to use them as lithium batteries to improve the charging performance of the battery.

Chapter 8
Application and Prospects

Graphene with its many unique properties has captured the attention of people in various fields. This is especially true after development of technologies that was able to produce graphene and related products on a large scale. Patent projects and applications involving graphene are growing exponentially year after year. The application research of graphene is expected to bring about a technological revolution, and then change our lives and the world we live in!

Let us start with a brief introduction to the applications where Graphene is used based on its main characteristics.

8.1 Energy Materials

Energy storage is one of the important areas of graphene applications. With the increasing demand for energy, it is necessary to develop energy sources that meet environmental protection standards, have large storage capacity, and can quickly charge and discharge. Graphene materials have ultra-high electron transport capabilities, giving them high power density and fast charging characteristics. Because the storage of electric energy is necessarily accompanied by charging and discharging, with the participation of graphene materials, the charging and discharging process will be faster. At present, typical applications of graphene in the field of energy storage include batteries and super capacitors.

The role of batteries in modern civilization is self-evident. In particular, they can be used as mobile power sources, including lithium-ion batteries in portable electronic products such as mobile phones that everyone cannot live without, and lead-acid batteries for electric vehicles that will become the main environmentally-friendly transportation tool. The principles of batteries can differ slightly but their essence is to convert chemical energy into electrical energy. Mobile power sources require multiple cycle's discharges and recharges, researchers hope that graphene can accelerate this process. Today, the so-called "graphene battery" currently promoted by

© Guangxi Science & Technology Publishing House 2022
T. Zhang, *Graphene*, https://doi.org/10.1007/978-981-16-4589-1_8

the media and some companies is misleading because, the battery is still the original lithium-ion battery or lead-acid battery with a certain amount of graphene material mixed in the electrode material to help improve the battery's conductivity. There are indeed a small number of scientific research teams working on pure graphene batteries or capacitors, but as of 2017, the claimed "graphene batteries" is not single-layer graphene, rather graphene powder, or multilayer graphene.

As an example, the media reported at the end of 2017 that "Samsung developed a graphene battery" actually means that they have developed a material called "graphene balls", that is, they have deposited on a silica substrate. Multi-layer graphene is grown to make graphene balls. Then a small amount of this material is used in a lithium battery as an electrode, thereby improving the battery's volumetric energy density and fast electrical conductivity.

Many experiments using graphene in batteries are still in the development stage with, research teams studying anode materials for lithium-ion batteries related to graphene.

The lithium-ion batteries that are widely used today are different from the earliest lithium batteries that used metallic lithium. Back then metal lithium has inherent instability and once caused safety problems for lithium batteries. Newer Lithium-ion batteries do not use metallic lithium, and are composed of positive and negative electrodes, separators and electrolyte. The safety of lithium-ion batteries can basically be guaranteed. Since Sony produced the first batch of lithium-ion batteries in 1991, it has become the most promising and fastest-growing market. From mobile phones, cameras, power tools, to Tesla cars, these batteries are widely used.

Lithium-ion batteries use lithium-containing oxides (both containing Li+) as the positive electrode (cathode), generally coke or graphite is used to form the negative electrode (anode), and the electrolyte is the conductor. When charging, lithium ions move from the positive electrode to the negative electrode through the electrolyte; when discharging, they move in the opposite direction.

The anode (or negative electrode) material of a lithium-ion battery is responsible for receiving lithium ions, which is vital to the performance of the battery and is the key to improving the performance of the lithium-ion battery. Therefore, researchers often try to replace graphite with other materials.

The University of California, Los Angeles has made a three-dimensional graphene porous structure. Using this material as the anode of a lithium-ion battery not only makes the penetration of lithium ions faster, but also inherits the large surface area and excellent conductivity, this improved the exchange and conductivity of lithium ions.

The South Korean scientific research team also invented a three-dimensional graphene material that can improve the performance of lithium-ion batteries. Compared with conventional lithium-ion batteries, the charging speed is faster and the capacity will not decrease.

Another idea is to replace graphite with "silicon-based" materials. In early 2018, a research team from the Department of Manufacturing Engineering (WMG) of the University of Warwick in the United Kingdom synthesized a lithium-ion battery anode material, called a silicon-high-quality thin-layer graphene (Si-FLG) composite

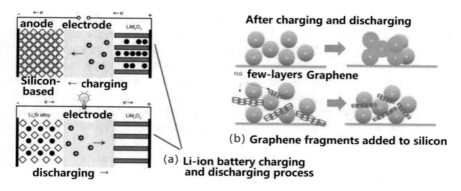

Fig. 8.1 Lithium-ion battery and graphene (https://scitechdaily.com/graphene-girders-could-dou ble-the-life-of-rechargeable-lithium-ion-batteries/)

electrode, and used it as lithium Alternative to graphite anode for ion batteries. The principle is to dope few sheet-like layers of graphene into the silicon-based anode, which will successfully and effectively form a separator between the silicon and the electrolyte, so that the battery maintains the separation between the silicon particles during each charging cycle, as shown in the Fig. 8.1. The use of this anode structure will greatly improve the cycle characteristics, electrode resistance and diffusion characteristics of the electrode, and extend the service life of the battery.

Graphene can make batteries that are lightweight, durable, suitable for large-capacity energy storage, shorten the charging time, maintain its charging capacity, and extend the life of the battery. These advantages are essential for electric vehicles.

In addition, due to the good optical properties of almost completely transparent graphene, it also shows broad application prospects in the solar cell industry. This transparent conductive film has a very wide spectral absorption range and high photo-electric conversion efficiency, and is suitable for manufacturing Solar battery. In 2017, MIT researchers developed a flexible and transparent graphene solar cell. It can be installed on all kinds of material surfaces, glass, plastic, paper, etc. It is expected that it will eventually be possible to realize a cheap solar cell covering a wide area, just like the printed newspaper of a newspaper printer, which is rolled into rolls and shipped to various places.

Graphene can conduct electricity and transmit light, both of which are good, making it an excellent application prospect in transparent conductivity electrodes. Organic photovoltaic cells, liquid crystal displays, organic light-emitting diodes, etc. all require good transparent conductive electrode materials. The commonly used conductive electrode material is indium tin oxide (ITO), which has high brittleness and is easily damaged, and its mechanical properties are not comparable to graphene. The use of graphene to replace ITO can be pricey, therefore, the preparation of large-area, continuous, transparent, and high-conductivity few-layer graphene films is very important.

Although there are certain types of batteries that can store a lot of energy, they are bulky, heavy, and release energy slowly. Supercapacitors can charge and discharge more quickly, but they have much less energy than batteries. The application of graphene in this field provides exciting new possibilities for energy storage, with high charging and discharging rates and even economic benefits. The improvement of graphene blurs the traditional difference between supercapacitors and batteries. The combined use of graphene batteries and supercapacitors can produce amazing results, such as improving the mileage and efficiency of electric vehicles.

8.2 Electronic Devices

The largest application of graphene is in the electronics industry. Some examples including graphene radio frequency tags (RFID), graphene electromagnetic interference shielding (EMI), graphene biosensors, gas and humidity sensors and other applications in electronic components. The electron mobility of graphene at room temperature exceeds 15,000 cm^2/(V s), which is higher than carbon nanotubes or silicon crystals, and its resistivity is only about 10^{-6} Ω cm, which is lower than copper or silver. With those exceptional properties that out performs silicon and copper, graphene is expected to be used to develop a new generation of thinner and faster conduction speed electronic components or transistors.

Because of its ultra-thin structure and excellent physical properties, graphene is expected to show attractive application prospects on FET (Field Effect Tube). Studies have found that graphene FET has a lower operating voltage, and its electron and hole mobility reach 5400 and 4400 cm^2/(V·s) respectively, which are much higher than traditional semiconductor materials such as SiC and Si. There is a fatal flaw though when using graphene to makes logic switch circuits. Graphene's conductivity at the Fermi level will not drop to zero like ordinary semiconductors, but reach a minimum value, which makes the graphene FET always "on" status.

About 90% of the devices that make up integrated circuit chips are derived from silicon-based CMOS, and the development of silicon-based CMOS technology will reach its performance limit in 2020. The reason is that as the size of transistors shrinks, the uniformity of device processing becomes more and more difficult. Using traditional microelectronic processing technology, the current processing accuracy is about 5 nm. With the continuous shrinking of device dimensions, the physical length of the corresponding transistor channel is only a dozen nanometers.

In 2016, as an important milestone was achieved in the integration of graphene in graphene photonics, researchers from the European Graphene Flagship demonstrated how graphene could provide a simple solution for silicon photo detection at telecommunication wavelengths, as shown in Fig. 8.2a.

Figure 8.2b shows the world's first multi-level graphene radio frequency receiver that IBM researchers claimed to be successful in 2014. They used the mainstream silicon CMOS process to manufacture this full-featured graphene integrated circuit, and successfully conducted a text message sending and receiving test. It is said that

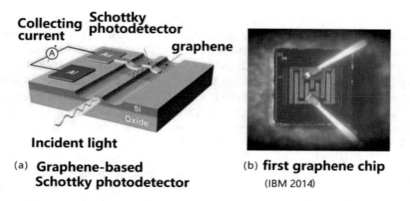

Incident light

(a) **Graphene-based**   (b) **first graphene chip**
Schottky photodetector   (IBM 2014)

Fig. 8.2 Graphene applied to electronic devices (http://www-g.eng.cam.ac.uk/nms/highlights. html)

the graphene receiver researched by IBM is composed of 3 transistors, 4 inductors, 2 capacitors and 2 resistors. The performance is 1000 times better than the previous graphene integrated circuits, and comparable to modern-day wireless communication with silicon technology.

Many large companies and research centers in the world have carried out the research and development of graphene semiconductor devices. Sungkyunkwan University in South Korea has developed a highly stable n-type graphene semiconductor. Columbia University has developed graphene-silicon photoelectric hybrid chips. IBM researchers have developed graphene field-effect transistors with excellent frequency performance, with a cut-off frequency of up to 100 GHz, which far exceeds the cut-off frequency of the most advanced silicon transistor: 40 GHz under the same gate length condition.

At present, the application research of graphene in electronic devices also includes conductive ink, heat sink, radio frequency identification, smart packaging, touch screen, sensor, etc.

Is graphene really comparable to silicon across the board? Only time will tell.

8.3   Ultra-thin Material that Soft and Rigid

In addition to specific electrical properties, graphene has its own unique features as an applied material. Aside from its super mechanical and thermal properties, graphene is also a completely transparent and flexible material that can be bent. There is always the vision that the computer screens of the future should not be in the current cold and rigid form but should be something that can be rolled up and carried around. Other application areas including e-books, electronic paper, flexible touch screens, smart fabrics, transparent mobile phones, curved mobile phones, flexible screens etc., the list can go on and on.

Being both conductive and transparent, graphene has the advantages of being used as a material for touch screen displays. The flexibility of being able to roll graphene is its biggest advantage and it is indeed exhilarating. The current ITO conductive glass is unmatched in these aspects, both conductivity and transparency ae not as good as graphene, and the biggest shortcoming is that it has no flexibility and breaks when it is bent. The disadvantage of graphene is the cost. At current rate it is too expensive and lacks market competitiveness. The ability to mass-product high-quality graphene is a must to enable the material being widely used in the communications market.

Wearable products are an emerging industry with huge potential in the upcoming market. The flexibility and rollability of graphene materials are ideal candidates. For example, wearable products for medical and healthcare purposes use electronic devices with multiple sensors to quickly transmit information such as human body temperature, pulse, blood pressure, etc. "Sensing" can be said to be the "good game" of graphene, because it is a two-dimensional grid that is light, thin, and strong, and has a large surface area. The grid is full of "bare" carbon atoms, it is easy to perceive any small changes in the surrounding environment, and even a gas molecule adsorption or release can be detected by sensitive graphene sensors. After acquiring the data, electrons can quickly transmit information to the receiver. Sensors based on graphene materials can detect electrical signals from various parts of the muscles of the human body to drive manipulators, etc., and can also be used on prostheses. In addition, any wearable product still needs a power source. It is possible to use graphene to make a flexible and bendable battery, which can be easily installed on the clothes worn by the monitored elderly. To make it even better those type of flexible battery can be laundered in the washing machine.

Extend beyond wearables, the graphene-based flexible Wi-Fi receiver is ideal for flexible electronic devices and biomedical equipment. Graphene can also play a part in home and office automation when it is possible to make ultra-thin and flexible electronic devices.

8.4   Lightweight and Super Strong Material

Graphene can be used as a lightweight and super strong material for biomedicine, transportation, and defense and aerospace. For example, graphene supercars tested by graphene researchers at the University of Manchester in collaboration with BAC automotive companies are eye-catching because automotive structural parts developed based on graphene are lighter and stronger than carbon fiber composites, which improve energy utilization. This kind of lightweight and high-strength graphene composite material can also be applied to aerospace fields where the weight and rigidity of the material are critical. In addition, the current car collision detection system based on graphene can work under visible light and infrared light, so it can avoid collisions under any weather conditions and is very useful for autonomous driving. Military uses include graphene bulletproof helmets, body armor and bulletproof armor.

As an all-carbon material, graphene has good biocompatibility and can be used as a drug carrier, or as an implant in implantable technology for smart therapy.

8.5 Environmental Purification

The pollution of heavy metals and other harmful substances to water sources is becoming more severe. Purifying water is a major event affecting the national economy and people's livelihood. Even if we do not talk about pollution, the shortage of water resources is a serious problem the world is facing. Statistics show that nearly two-thirds of the world's population will face water shortages in the coming decades. The problem of water purification has always been a hot topic in scientific research.

Activated carbon is often widely used in the purification process of chemical industry, electronics, medicine, food, domestic and industrial water, because it has a porous solid surface that can adsorb and remove organic or toxic substances to purify the water. Carbon element graphene should have greater advantages in this kind of adsorption and purification process, especially when an ideal single-layer graphene is not needed. Graphene oxide (GO) is easier to mass produce, which makes it is more attractive, and now the research in this area has begun to bear fruit.

Graphene has a unique two-dimensional structure and pore size distribution, a relatively large specific surface area, and the properties of the surface can be modified. Graphene has good metal ion adsorption performance, simple adsorption characteristics, and high efficiency. A single-layer graphene has no active groups on the surface and can only adsorb heavy metal ions through van der Waals forces. Graphene oxide derivatives have a large number of oxygen-containing functional groups on the surface, which are negatively charged in water and easily adsorb most heavy metal cations. The structure of graphene oxide can be further improved to enhance electrostatic attraction, and form new compounds with better adsorption effects. It has important research value and application prospects in the adsorption of heavy metal ions.

The penetration by material size can also be used to purify water. Original graphene is very dense and impermeable, because the electron cloud formed by its π orbital blocks the voids in the ring, making it impossible for He molecules with a small radius to pass. Scientists have since then found that the interaction of graphene and water is somewhat puzzling. The graphene film that appears to be repellent to water can at times form many capillary channels under certain conditions, allowing water to penetrate quickly. In addition, researchers have also adopted the method of punching sub-nanometer-level holes in the graphene film to form filterable sheets. The size of the pores can be controlled and adjusted in advance, such as designing their size to only allow water molecules to pass through, and block other larger salt molecules, heavy metal impurity molecules, etc., to achieve the purpose of purifying water, as shown in Fig. 8.3.

As mentioned earlier, the earth's water resources are facing a crisis of shortage, in general, there is no shortage of water on the earth, they are just not drinkable.

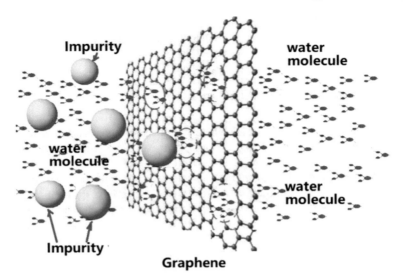

Fig. 8.3 Graphene-based materials used for water purification

Seawater accounts for 97% of the earth's water sources, finding a convenient and cheap way to filter the high salinity of seawater into drinking water will solve the water crisis.

Graphene-related products may come in handy. In early 2018, researchers from the Australian Commonwealth Science and Industry Organization (CSIRO) developed a new type of graphene filter membrane made from renewable soybean oil. They named this special material based on graphene Graphair. It is said to be a thin layer of pure carbon composed of micro-nanotubes. The materials with unique atomic structure that they researched and developed nano-channels only allow water molecules to pass through, eliminating salt and other larger pollutant particles.

The new technology is very efficient and can directly filter the water samples collected in Sydney to be drinkable without further processing. The new material Graphair membrane is simple, cheap, and easy to manufacture. It is with high hope this will solve the problem of drinkable water crisis.

8.6 Biomedicine

Graphene can be used in biomedicine-related fields such as bacterial detection and diagnostic devices. Chinese scientists have found that graphene oxide is very effective in inhibiting the growth of *E. coli*, it is possible to use graphene as an antibacterial substance in medical devices or food packaging.

There are a lot of studies trying to apply graphene in biomedicine, let us use DNA sequencing as an example.

In basic biological research and application, from disease diagnosis, drug development, forensic identification to anthropological research, knowledge of DNA sequence has become an indispensable instrument. DNA sequencing can be used to determine the sequence of a single gene of any organism (including humans and other animals, plants, and microorganisms), and it is also the most effective method for sequencing RNA or protein. In molecular biology, the information obtained by DNA sequencing identifies the genetic changes caused by diseases, which can help determine potential drug targets; in the study of human evolution, DNA sequencing is used to determine the correlation between different races and how the entire human race has evolved and developed.

The purpose of sequencing is to analyze and determine the base sequence of DNA fragments, that is, the arrangement of adenine (A), thymine (T), cytosine (C) and guanine (G). Fast DNA sequencing methods will greatly promote the research of biology, medicine and pharmacology.

DNA sequencing using graphene is an extension of the original nanopore sequencing principle. Nanopore sequencing relies on charged particles (ions) passing through nanopores to trigger potential changes to detect base sequences. Based on the high sensitivity of graphene, scientists thought of using graphene flakes as a sensor for sequencing methods.

Make a nano-hole about the width of DNA on the graphene sheet, and let the DNA strand pass through the nano-hole, as shown in Fig. 8.4. When the base passes near the graphene nano-hole, the generated mechanical strain signal will affect the conductivity of the graphene and cause a potential change. The four different bases (A, C, G, T) will have different effects on the conductivity of graphene. By detecting and amplifying the tiny voltage difference generated when DNA molecules pass through with appropriate circuitry, it is possible to know which base is swimming through the nano-hole.

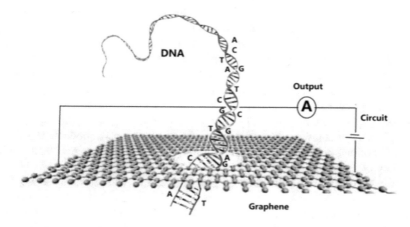

Fig. 8.4 Graphene used for DNA sequencing

Experts believe that graphene nano-hole DNA sequencing is a highly accurate and efficient method. The simulation experiment results show that the sequencing method can identify 66 billion bases per second, with an accuracy of 90% and no false positives; it is faster and cheaper than traditional sequencing methods.

8.7 Graphene and Glass

Since the discovery of graphene in 2004, all relevant scientific research units and companies in China have shown great success in the preparation and application of graphene. China has become one of the most active countries in graphene research and application development, and the number of papers published and patent applications related to graphene ranks among the top in the world. According to statistics, as of August 29, 2018, the total number of graphene-related patent applications worldwide reached 51,054, while the total number of patents from China was 33,987, accounting for 66.57%, far ahead of the second-ranked United States (7652) And South Korea ranked third. (7015).

Many universities, research institutes, and companies in China are full of enthusiasm for the research and development of graphene products, and have made many world-leading results in the basic research, preparation technology and application of graphene. The following is just an example of applied research.

The graphene research team of Peking University has developed chemical vapor deposition (CVD), successfully prepared graphene on glass, produced super graphene glass, and developed many application projects [25]. A more compelling product developed directly from this is the super graphene optical fiber, because the optical fiber is originally an ultra-fine silica glass fiber. The preparation method they developed can prepare one to several layers of graphene on the outer surface of the optical fiber or the inner wall of the optical fiber, and obtain good layer number controllability, which is the first international breakthrough.

The use of graphene fiber can be widespread. Optical fiber is an indispensable material in the modern communication field, compared with traditional optical fiber; graphene optical fiber has a long life and high mechanical strength. In addition, it is possible to use the excellent conductivity of graphene to realize the integration of cables and optical cables.

Super graphene fiber can also be used for fiber optic sensors, fiber optic endoscopes, fiber optic lighting, etc., such as a new type of biosensor shown in Fig. 8.5. A circle of graphene around the fiber part of the sensing area can enhance the detection sensitivity of the sensor.

As we have learned in this chapter, graphene can be used in many areas, from theoretical research to engineering applications; the attractiveness of this material has not diminished. There are high-end and low-end graphene, especially the broad and multi-layer graphene, which is easy to prepare, low cost, and can be widely used.

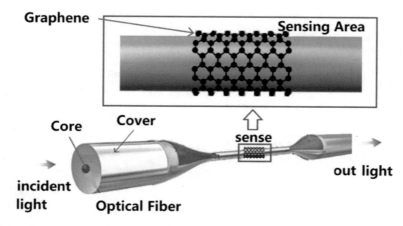

Fig. 8.5   Graphene is used for optical fiber sensor

References

1. L.D. Landau, E.M. Lifshitz, *Statistical Physics, Part I* (Pergamon, Oxford, 1980), pp. 137–138
2. K.S. Novoselov, A.K. Geim, S.V. Morozov, D. Jiang, Y. Zhang, S.V. Dubonos, I.V. Grigorieva, A.A. Firsov, Electric field effect in atomically thin carbon films. Science 306(5696), 666–669 (2004)
3. Andre Geim—Nobel Lecture: Random Walk to Graphene, https://www.nobelprize.org/upl oads/2018/06/geim_lecture.pdf
4. M.V. Berry, A.K. Geim, Of flying frogs and levitrons (PDF). Eur. J. Phys. **18**(4), 307–313 (1997)
5. A.K. Geim和H.A.M.S. ter Tisha, Physica B 294–295, 736–739 (2001)
6. M. Planck, On the theory of the energy distribution law of the normal. Spectrum. Verhandl. Dtsch. Phys. Ges., 2, 237, 1900
7. Concerning an heuristic point of view toward the emission and transformation of light. Annalen der Physik 17 (1905), 132–148. http://einsteinpapers.press.princeton.edu/vol2-trans/100
8. N. Bohr, On the constitution of atoms and molecules. Philos. Magazine 26, 1–25, 476–502, 857–875 (1913)
9. Recherches sur la théorie des quanta (Researches on the quantum theory), Thesis, Paris (1924)
10. E. Schrödinger, An undulatory theory of the mechanics of atoms and molecules. Phys. Rev. **28**(6), 1049–1070 (1926)
11. A. Einstein, B. Podolsky, N. Rosen, Can quantum mechanics description of physical reality be considered complete? Phys. Rev. 47, 777
12. D.J. Griffiths, *Introduction to Quantum Mechanics* (1995). Prentice Hall. ISBN 0-13-124405-1
13. I.M. Mikhailovskij, E.V. Sadanov, T.I. Mazilova, V.A. Ksenofontov, O.A. Velicodnaja, Imaging the atomic orbitals of carbon atomic chains with field-emission electron microscopy. Phys. Rev. B **80** (2009)
14. A.S. Stodolna, A. Rouzée, F. Lépine, S. Cohen, F. Robicheaux, A. Gijsbertsen, J.H. Jungmann,C. Bordas, M.J.J. Vrakking, Hydrogen atoms under magnification: directobservation of the nodal structure of stark states. Phys. Rev. Lett. **110**(21) (2013). https://doi.org/10.1103/Phy sRevLett.110.213001
15. A.W. Robertson, J.H. Warner, Atomic resolution imaging of graphene by transmission electron microscopy. Nanoscale **5**, 4079–4093 (2013). http://pubs.rsc.org/en/content/articlelanding/2013/nr/c3nr00934c#!divAbstract
16. L. Pardini et al., Mapping atomic orbitals with the transmission electron microscope: images of defective graphene predicted from first-principles theory. Phys. Rev. Lett. (2016). https://doi.org/10.1103/PhysRevLett.117.036801
17. M. Born, K. Huang, *Dynamical Theory of Crystal Lattices*, Oxford University Press, USA (November 5, 1998), ISBN-13 : 978-0198503699
18. K.S. Novoselov, A.K. Geim, S.V. Morozov et al., Two-dimensional gas of massless Dirac fermions in graphene. Nature **438**(7065), 197–200 (2005)

19. Y. Zhang, Y.-W. Tan, H.L. Stormer, P. Kim, Nature **438**, 201 (2005)
20. K.S. Novoselov, Z. Jiang, Y. Zhang, S.V. Morozov, H.L. Stormer, U. Zeitler, J.C. Maan, G.S. Boebinger, P. Kim, A.K. Geim, Room-temperature quantum hall effect in graphene. Science **315**, 1379 (2007)
21. M.P. Lavin-Lopez, L. Sanchez-Silva, J.L. Valverde, A. Romero, CVD-graphene growth on different polycrystalline transition metals. AIMS Mater. Sci. **4**(1), 194–208 (2017). https://doi.org/10.3934/matersci.2017.1.194
22. Y. Cao, P. Jarillo-Herrero et al., Correlated insulator behaviour at half-filling in magic-angle graphene superlattices. Nature (2018)
23. Y. Cao, P. Jarillo-Herrero et al., Unconventional superconductivity in magic-angle graphene superlattices. Nature (2018)
24. E.J. Mele, Novel electronic states seen in graphene. Nature (2018)
25. J. Sun, Y. Chen, X. Cai, B. Ma, Z. Chen, M.K. Priydarshi, K. Chen, T. Gao, X. Song, Q. Ji, X. Guo, D. Zou, Y. Zhang, Z. Liu, Direct low-temperature synthesis of graphene on various glasses by plasma-enhanced chemical vapor deposition for versatile, cost-effective electrodes. Nano Res. **8**(11), 3496–3504 (2015)

Printed in the United States
by Baker & Taylor Publisher Services